EINSTEIN'S SPACE
AND VAN GOGH'S SKY

Other books by LAWRENCE LeSHAN

You Can Fight for Your Life:
    Emotional Factors in the Causation of Cancer
Alternate Realities
How to Meditate
The Medium, the Mystic, & the Physicist

Other books by HENRY MARGENAU

Physics of Philosophy, Selected Essays
Ethics of Science
The Scientist
Open Vistas
Nature of Physical Reality
Mathematics of Physics and Chemistry
Foundations of Physics

EINSTEIN'S
AND VAN

SPACE
GOGH'S SKY

Physical Reality and Beyond

**by LAWRENCE LeSHAN
and HENRY MARGENAU**

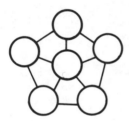

COLLIER BOOKS
Macmillan Publishing Company
New York

Macmillan Publishing Company
866 Third Avenue, New York, N.Y. 10022
Collier Macmillan Canada, Inc.

Library of Congress Cataloging in Publication Data
LeShan, Lawrence L., 1920–
    Einstein's space and Van Gogh's sky.
    Includes bibliographical references and index.
    1. Science—Philosophy.   2. Science—Methodology.
3. Reality.   4. Reductionism.   I. Margenau, Henry,
1901–   . II. Title.
Q175.L444   1983   501   83–7356
ISBN 0–02–093180–8

First Collier Books Edition 1983

10   9   8   7   6   5   4   3   2   1

Einstein's Space and Van Gogh's Sky is also published
in a hardcover edition by Macmillan Publishing Company.

Printed in the United States of America

Dedicated to Arthur Twitchell,
who generously sponsored
the Conference on Human Potentialities
at Palma, Majorca,
which brought the authors together
and made this work possible.

Contents

Preface

THIS BOOK HAS GROWN from two others, *The Nature of Physical Reality* (Margenau) *Alternate Realities* (LeShan). The first deals in solid scientific fashion with sensory or physical reality, basing its philosophic arguments in customary, established fashion on the physical science of its day. The second presents queries and some unorthodox suggestions and answers concerning experiences that lie largely outside the interest of scientists—indeed, in some respects defy acceptance by them. The styles of the two books conform to their purposes: one is conventional, precise, and (somewhat) technical, limited in its scope; the other is unconventional, expansive, probing into the obscure, and daring in its metaphors and quotations.

We met at a conference devoted to some important philosophic problems of modern psychology. On this occasion we discussed our books, and two things became clear: The first-named book takes "consciousness" for granted, shows how the mind uses perception and reason to construe reality from its own experiences. The second emphasizes that the mind has numerous states or phases that transcend the processes and experiences upon which sensory or physical reality is built, but that we possess no universally accepted methodology, no single method of approach that gives them a solid status. On comparing the diverse approaches to, and functions of, the mind demonstrated by those books, we decided to undertake a wide-searching study to attempt to recognize and establish validity for at least some nonsensory experiences.

This meant that an account had to be given not only of perceptions and the reasoning they provoke, as in *The Nature of Physical*

Reality, but also of all other mental states, states vaguely referred to as inner feelings, hopes, intentions, expectations, memories, pain, decisions. These states have in common one feature that distinguishes them from the percepts that lead to physical reality: They cannot be quantified, cannot be measured in the direct physical sense. We ask the question: How then can we deal with these percepts? Is there a procedure allowing us to deal with them in a systematic manner that might be said to impart to them "reality"?

This book is meant to clear the way to this goal. We hope that it can be followed by a more incisive one on the nature of the mind in which the goal is brought within scientific reach, in which consciousness is objectively analyzed—not merely in terms of its attendant behavior, but intrinsically.

A physicist interested in the systematics of reality and a psychologist interested in alternate methods of construing or organizing reality would appear to be aiming at different goals. But we found that there were not two goals, but rather two interacting and mutually supporting methods for investigating the same goal—an understanding of the relationships between consciousness and reality.

Each culture organizes its reality in a specific way, and its members are convinced that this is the only correct view of the universe. This organization was developed in response to the most pressing and overwhelming problems of the culture's previous period and is designed to answer these problems.

Our present Western culture developed its way of organizing reality in response to the need to control disease (after the Black Plague) and to other problems relating to the control of the environment. This method is characterized by two central ideas: materialism ("What I can see and touch is most real, and everything else is less real") and Cartesian dualism ("This is me [my consciousness] in here, and the rest of the world is out there"). For the needs that gave it birth it is a very powerful method and has given us immense and unprecedented control over the "outside" environment. However, as do all organizations of reality, it has built into it inexorable problems that it is not equipped to solve. These are the major problems our culture is facing today, and it is these, as the historian Arthur Toynbee has shown, that we must solve or else we will go under.

Chief among these problems is the fact that this method is a strong tool for studying half of the Cartesian dualism (". . . the rest of the world out there"), and a very weak, inadequate tool for studying the other half ("This is me in here . . ."). We inevitably arrive at a situation in which we increase our understanding of matter and energy but do not increase our understanding of the minds that direct the use of matter and energy. To give one obvious example, we have greatly increased our ability to wage war since Hellenistic times, but we have not increased our understanding of the causes of war. In order to solve problems such as this which arose from the development of our present structure of knowledge, a new organization of reality is crucial.

It is the central thesis of this book that this new method is already in development and service in a number of scientific fields and that a clear statement of it will accelerate its use in the solving of our most pressing problems. We attempt to show here what that method is, how it has developed in two areas (physics and psychology), and to demonstrate that it *is* organized in such a way as to make possible the development of new solutions to some of the problems of our culture. The significance of this choice is of course more embracive than the two words "physics" and "psychology" imply: It distinguishes the sciences of the external world from those of the mind. The latter include sociology, some aspects of philosophy, the arts, ethics, and even religion.

We begin with an introduction to the new view of reality, particularly as it has been developed in physics and psychology. In the first section of this book we present this new concept in a general and introductory manner. In the second section we explore the methodology of the physical scientist and look in depth at the problem of reductionism—the idea that some aspects of reality are more "real" than others, that some can be validly "reduced" to others. Our appraisal of reductionism, our rejection of its customary materialistic form based on recent discoveries, forms the central part of the book. Contrary to the usual efforts in this direction, we have found that in physics (let alone elsewhere) reductionism does not work. Biology cannot usefully be considered "nothing but" chemistry, and human behavior cannot usefully be considered "nothing but" combinations of reflexes. Within physics itself the use of reductionism has produced more of a shambles than it has produced progress. In this section we explore some of the implications of present-day science in such diverse areas as "reality in the microcosm and the macrocosm," "causality," and "purpose."

In the third section we apply the new theory to five specific areas in which the problems have not yielded to the "old" methodology. These are the areas of interest to the social scientist; they concern the artist, the musician, and the parapsychologist; they refer to the problem of ethics and to the problem of consciousness in general. In each of these this application leads, it seems to us, to fruitful conclusions.

The collaboration between a psychologist and a physicist, though both are interested in the philosophical implications of their subjects, can hardly be expected to result in a perfectly uniform style. We therefore request the reader to be tolerant of differences in our manner of presentation.

The Meanings of Reality

THIS FIRST SECTION aims to deprive, not to say rescue, the word reality of its ingrained, fixed, and limited meaning. It will disappoint, or perhaps enlighten, the reader who is convinced that only objects that can be sensed are real. The term should be revised for two basic reasons. One must be seen in some of the recent discoveries in the physical sciences and in psychology. The works of Heisenberg, Schrödinger, Einstein, Born, Freud, and Jung cannot be understood in terms of the common hear-see-touch methods of defining what is real. Inferences, carefully conducted by rules and conventions that are highly specific, often serve that purpose.

Too, a revision of our ideas about reality is required by certain subjects that the hard-boiled materialist and the eager experimentalist who spurn an appeal to theory term "pseudo-sciences." These include to a large extent sociology and economics, clearly religion, and especially a very old subject that has recently attained increasing popular as well as scientific interest: parapsychology. The old attitude, which shrugged off the claims made by researchers in this latter field by attributing them to fraud, chance, and ineptitude on the part of the experimenters, can no longer be maintained. To reach this conclusion has not been easy for one of us in particular, who was accustomed to the rigors of physical research. But a careful study of a limited area of parapsychology (extrasensory perception and certain experiments in telekinesis) has convinced him that the care taken by some workers in these fields, their understanding of the theories of error, statistics, and probability, are at least as thorough as that of most of his scientific colleagues. To be sure, the literature in this area is extensive, and much of it is subject to criticism. In some journals editorial discrimination is not as strict as in the established sciences, but the disciplined investigations can no longer be ignored.

In the following two chapers we survey the meaning of reality as presently construed by our culture, present and name its major features, and then widen its sense to allow inclusion of what seems acceptable in the pseudosciences.

1 Alternate Realities

"... Einstein's space is no closer to reality than Van Gogh's sky. The glory of science is not in a truth more absolute than the truth of Bach or Tolstoy, but in the act of creation itself. The scientist's discoveries impose his own order on chaos, as the composer or painter imposes his; an order that always refers to limited aspects of reality, and is based on the observer's frame of reference, which differs from period to period as a Rembrandt nude differs from a nude by Manet."
—ARTHUR KOESTLER[1]

THE WORD "REALITY," as used in ordinary discourse, has a definite, easily comprehensible, and ultimate meaning. New phenomena often fall prey to its menacing stare. This narrow definition, a product of our past, is now badly hampering our progress.

Each individual is born into a culture, and its orientations and basic beliefs shape him and remain deeply rooted in his personality all of his life. If he moves into a new culture with other orientations and basic beliefs, the two versions of reality are dissonant within him. Even after he is a firmly functioning member of the new culture, the orientations of his beginnings still influence him.

As it is with an individual, so it is with a field of knowledge. The sources from which a field grew remain within it as a shadow skeleton, and they partly define for it what is real and what is true, what is sense and what is nonsense—in short, what is the basic shape or essence of reality. When the field develops so that new data contradict these old beliefs, a basic conflict develops in the field of knowledge. There is great difficulty and struggle in recognizing, organizing, and solving the new problems presented by the conflict of the new data and the old beliefs and basic orientations. In the struggle confusions arise, and there is a loss of communication among many of the students of the field of knowledge. Today science is in the midst of such a struggle. Some of the basic assumptions, that shadow skeleton of our way of organizing experience, are being contradicted by data emerging in a variety of scientific fields.

Science grew up in the 17th and 18th centuries, a time when the primary world view was that the cosmos had been made by one God,

[3]

a rational God. This was seen as common sense. The cosmos was therefore rational, and there was only one meaning to the term *rational*. The task of science was to understand the rational structure of the universe. Petrarch, dealing at the beginning of the Renaissance with the problem of how to develop a scientific orientation in a religiously oriented culture, wrote that one way to worship God was to comprehend and therefore admire His handiwork. In this view, which was the view of the culture, all things, having been made by one rational God, were made in the same way.[2] That there is one rationality governing the entire cosmos is now one of the most basic beliefs and articles of faith of science.[3] Whoever doubts it is seen not as a scientist, but as a superstitious heretic.

Built in to the growing field of science and our structure of knowledge, then, are the assumptions that the world is rational and that there is only one meaning to that term; that the world is consistent in its rationality, that all phenomena in it can be understood in consistent terms and follow consistent laws that are accessible to reason. There is one rationality, and everything from atoms to galaxies, from dreams to machines, from human behavior to the stroke of lightning in the sky, can be understood in its terms. Deepening and widening this single understanding is the work of science.

With a rational God and a rational cosmos, there can be no room for exceptions from the laws of reality. Every entity follows the laws of a single rationality, and every occurrence expresses these laws. "There is no such thing," wrote St. Augustine, "as a miracle which violates natural law. There are only occurrences which violate our limited knowledge of natural law." God himself was bound in the web of rationality.[4]

Gradually in the history of science this concept of the one rationality was clarified, and some of the basic laws of the concept emerged. The first scientific progress was made in that realm of experience in which things could be seen and touched. In this realm things could be, at least theoretically, separated from each other, counted, added, or subtracted, and so it seemed obvious from this one rationality (as there had been one God) that *since part of the universe was quantitative, then everything in the universe was quantitative.* (The meaning and significance of these terms and ideas will be clarified in chapters 3–6.) From this it was concluded that a field of science only progresses to the degree that it makes its data (the observables in the domain it had chosen for study) quantitative. This became a basic faith of science. So strong was this belief that people failed to realize that quantification (counting and measuring) is a human activity imposed on our knowledge of reality. We considered it to be part of reality itself. "God is a mathematician"—Leibniz's famous statement—expressed clearly our viewpoint and its history.[5]

Other specific ideas about the rational world in which everything could be seen and touched, gradually emerged. A major idea, apparently obvious from the data studied, involved the concepts of cause and effect: All events have causes, and these existed before the event. There is no such thing as an uncaused event, and causes come first in time followed by the event. In modern parlance the "state of a system" determines a later state. The past leads inexorably to the present. To speak of the future influencing an event in the present would be as nonsensical as to speak of an uncaused event. This idea is more complex than it seems at first thought. Aristotle, probing deeply into it, found it necessary to divide "cause" into four classes, each delineating one aspect of it. But with science making great strides in the realm of the visible and touchable—the "sensory" realm, a realm of billiard balls, of cogs, pistons, and gears, of levers and pulleys, of hammer and nail, of exploding gunpowder and flying cannonballs, a realm where the immediate cause and event could generally be quickly seen or, if not, almost invariably found on examination[6]—it became a basic belief of science that the past was the complete cause of the present and that all action was the result of interplay of forces from the past. The future had no power to shape the present. Not yet existing in the sensory realm, it had no power to pull events into existence: rather it was only a receptacle by means of which the past poured its contents into the present. (We shall analyze the meanings of cause and effect in detail in chapter 11.)

A clear implication of the assumption that the past causes the present is that the cosmos is predictable. If we know enough about the present, completely know the "state of the system," we can predict what will happen next. Since it was believed that cause and effect rule supreme and govern completely all events—an earthquake as well as the patterns of the tide upon the sand, the writing of the *Ninth Symphony* as well as the fall of the leaf of the tree, the swimming upstream of the salmon as well as the painting of the Mona Lisa—everything could be predicted in advance detail and precisely if we had sufficient exact knowledge. The 18th-century philosopher Laplace's great "intelligence," who knew the position and velocity of every atom in the universe and therefore could predict all future events forever, hovered in the background of our organization of knowledge as a perfectly realistic concept.[7]

Although this implication has had to be abandoned in the realm of quantum mechanics,[8] it has not been clearly understood that this abandonment means the full and complete collapse of the system of one rationality ruling the entire universe. A universe that is absolutely and completely predictable (at least theoretically) in some realms and not in others is not a universe that is run in all of its aspects by the same laws. (Of course there *are* certain either/or situations. For example, one is pregnant or not.) A completely consistent cosmos cannot be inconsistent in one area. One exception collapses it all.[9]

Based also on the structure of the realm in which science made its first great advances, a third assumption about the one rationality that governed the universe gradually developed. This, probably the last to emerge and the first to collapse, was that, as in the see-touch world, everything in the universe could be explained along mechanical, push-pull lines. The cosmos itself was a giant clockwork—possibly, as for Descartes, wound up and supervised by God—that could be truthfully explained by a mechanical model (and only validly explained by a mechanical model). This assumption, more subtle than it looks on first examination, includes the idea that everything that is really understood can be visualized, can be pictured by valid analogy. This *is* true of the realm of experience within the see-touch limits. It was not understood for a long time that it might not be equally true in all other realms—for instance, in the realm of things too small to be seen or touched even theoretically, or in the realm of consciousness. Gradually scientists came to realize that one simply could not visualize the observables in the quantum mechanical range (the "microcosm") in any consistent manner. An electron is a theoretical construct.[10] The idea of an electron without some of the numbers that define an ordinary particle is nonsense. There is no such concept, for example, as that of an electron having a fixed position at rest. (We will explore this in detail in chapter 7.) Due to our earlier orientation we had given our observables in this area the qualities of objects in the see-touch realm. But an electron is not a particle and not a wave, and calling it such only confounds our understanding. It is for this reason that one of us has named them *onta*, from the Greek for "beings."[11] We need a term like this to help keep clear that they are not "objects" in the ordinary sense but rather something that is fundamentally "other." (The singular for this word is *on*. To distinguish it from the preposition, this word will appear in italics whenever it is used in this book.)

It first became clear to scientists that the assumption that all phenomena could be visualized and explained by mechanical models was not completely valid through the work of James Clerk Maxwell and the general development of the concept of fields in physics.[12]

Although it was realized that this assumption did not hold true in various realms studied by physics, this discovery (like the discovery of the necessity of abandoning specific predictability in quantum mechanics) has remained curiously localized in science. In other scientific disciplines the belief is often still strong that what prevents us from making a useful and fruitful mechanical model of our data is our lack of knowledge rather than something inherent in the data. Physicists may recognize that it is invalid to conceptualize an electron as anything but a set of numbers (not, for example, as a small, round, rapidly spinning ball), but most psychologists generally still hold to the belief that someday, somehow, we will have mechanical models of the human mind and of human societies. This hope and assumption lay behind the brilliance and the deep searching of Freud, and the psychoanalytic system of describing personality might be viewed as the greatest monument ever built to this belief.

Science is struggling today with a profound problem involving this seeming paradox. Against the belief that everything that is, is real in the same sense and follows consistent laws, is pitted the knowledge that many data, including those of our inner experience, cannot be fitted into the same rational system that describes so well what is and what happens in the see-touch realm of experience. We have indicated briefly some of the ways this conflict has been dealt with in physics (as in the giving up of the mechanical model and in the abandonment of specific predictability of events in quantum mechanics and its replacement by statistical prediction) and will refer to these in greater detail later. For now, however, let us examine the problem from the viewpoint of psychology.

We have described three aspects of the one rationality believed to apply to the cosmos—quantification, cause and effect with resulting predictability, and the valid and necessary use of mechanical models. A great amount of serious work on the application of all three in the field of psychology has been unsuccessful. For example, let us look at the concept of quantification. Its precise meaning will be developed in following chapters, but the simple definition in general use will suffice here.

When you examine the history of psychology, you must see that serious attempts over the last hundred years to quantify inner experiences have failed. Indeed, so great has been this failure, so little hope held out for success, that today Behaviorists, a large group in psychology who try to deal with all behavior and experience on mechanical grounds, advocate ignoring the primary data of our field and our existence—our inner experience—in effect, pretending that it does not exist. A science that redefines itself in order to run away from its basic data is in a serious state indeed.

All psychologists remember the great efforts made to quantify inner life. There were "psychophysics" and the "taste tetrahedra" and the "smell prisms."[13] There was Herbart with his mathematics of the unconscious, Kurt Lewin with his Topological Psychology, and Clark Hull, Spence, Guthrie, and many others. In spite of their and other efforts, it has proved impossible to lay a yardstick against a fear or put a hope into a balance scale. I could say, "This table is precisely the same length as that one," but not, "Your pleasure is precisely as great as mine." And did the man who would walk a mile for a Camel have half the desire for a cigarette of the man who would walk two miles? With animals and people psychologists tried to find the equations connecting the qualitative inner experience with quantitative outer behavior, and they got nowhere. Attempt after attempt was made to quantify the inner experience, and they all turned to nothing. The best we could do with our experience involved qualitative determinations, and these we regarded as second-rate, as failures of science. We could say, "I have more pain than I had yesterday." When we tried to say, "I have nine dols of pain," we found we were talking nonsense. We could say that Rembrandt was a greater painter than Kandinsky. We could not say that Rembrandt was three and a half times better than Kandinsky. But so deep was the assumption that the whole universe was quantitative that we regarded this as a failure in our science rather than as a difference in the data themselves.

Our many attempts to apply aspects of the one-rationality hypothesis to human experience—to make it conform to reality as we observed it in the see-touch realm—also failed. Predictability was never the same as it is with billiard balls. After over seventy years of experience most psychotherapists have come to the conclusion that the only approach that makes sense when dealing with human behavior is that the past was determined and the future is free. (As we shall see in chapter 12, this is precisely the conclusion reached by a rigorous application of modern scientific methods to the data of the realm of consciousness and of the realm of meaningful behavior.) Since predictability rests on the idea of the past being the complete determinant of the present, however, a major problem existed. Central to our experience is the sense of *purpose* of our actions, of what we want for the future determining what we feel and do now. This is clearly observable. To ignore it is to ignore part of the basic data of our existence. To accept it is to destroy the one consistency which we feel governs both our inner experience and the behavior of billiard balls. To rant against "teleology" (the belief that the *goal* influences behavior; it does not do this with the movement of billiard balls or flying arrows, and was therefore often considered invalid when used as an explanatory factor in human behavior) as being "unscientific" is no answer. We *know* there is a difference when we reach to pick up a telephone as to whether our purpose is to phone a hospital to inquire after a sick child, to whisper love terms to one we are enamored of, to complain to a contractor that the promised work is behind schedule, or to make an obscene telephone call. To ignore these differences in purpose is nonsense. Yet, to accept them is to violate the basic axiom of modern science that the universe is consistent since there is no purpose involved when one billiard ball strikes another and each moves off with a speed and direction fully determined by the past.

Quantification, determinism, the attempt to build mechanical models, all failed in spite of serious and long-continuing effort. There remained only the faith that the one rationality, originally attributed to God the Creator, would some day be applied to our inner life by some new insight. What we ignored in maintaining this faith was the fact that it ran directly counter to all of our experience. We neither lived nor acted nor felt as if it were true. We lived in a number of completely different models of the universe, different ways of construing reality, different definitions of what was real and unreal, sensible and nonsensical, during the course of any one day. Let us illustrate this by a day in the life of an imaginary, hard-boiled, down-to-earth businessman.

In this man's everyday work, as he sits at his desk, he lives in a reality we all know very well. It's the reality we in the West ordinarily think of as the real one. It is the reality in which we tie our shoelaces and design the shoes, in which we buy airplane tickets and take a taxi to the airport. The businessman would say, as would most of us, that this is the only *real* reality, and every other one is actually some aberration or other, usually temporary.

One day the businessman comes home after work. He knows there has been some meningitis in the area, and he's worried about his three-year-old child. Sitting downstairs in the evening, he hears the child crying upstairs. As he goes upstairs, he is terribly frightened. He finds himself pleading, "Please, don't let it be meningitis." He is really praying. His whole consciousness is involved in this action. He is completely organized in such a way that this is the only thing that makes sense to him, that what he is doing at that point is the reasonable action to be doing. He does not question it. At that moment he is perceiving and reacting differently than he does during the rest of the day. At work he knows there would be absolutely no point in such pleading. The universe as he ordinarily construes it does not respond to emotion and prayer.

He arrives upstairs and finds to his vast relief that the child is not ill. The child has only awakened in the night upset and frightened. He soothes the child. He holds the child in his arms and says, "It's all right." What is really happening here? The child has awakened, confused and frightened, and the businessman reassures him by saying, "It's all right. The universe is friendly. Things are all right." Now in his ordinary, everyday state of consciousness, the way he ordinarily organizes reality, this is certainly not true. He lives in a world that eventually will kill both the businessman and the child and totally annihilate them both. One cannot say this to a child and also say, "It's all right, the universe is friendly." But the businessman is not lying. At this moment he is in a completely different reality than he was during the day or when he was coming up the stairs. Out of a deep sincerity he is saying, in effect, There is a way of being in the universe where love transcends death and where the cosmos will not annihilate us. Again, he has organized reality in a different way. And this is the way, at this moment, he perceives and reacts to reality, and this is what he *knows* at this moment to be the complete truth.[14]

After reassuring the child, the businessman comes downstairs. That evening he and his wife go dancing. During the evening he is dancing in his usual way, enjoying it more or less, thinking of various things, the music, his partner, what they'd been talking about, other people, and so forth. Suddenly he realizes that for a period of time—he is not sure exactly how long—everything was different. During this period of time that just passed he wasn't thinking about anything. He was not in a daze. He was not in a trance. He was not asleep. As a matter of fact he was very wide awake and alert, but his whole being was doing just one thing—dancing. After it was over he felt good, "charged up," slightly "high," and very pleasantly relaxed. If the period that passed is analyzed carefully, we find that again he had organized reality in a different way. No longer was he listening *to* the music, dancing *with* his wife, *avoiding* the other people, but rather he and the music and his partner were in a very fundamental sense—one. He was moving as if he were part of a network that included the music, the floor, the other people, and the whole scene. He was dancing far better than he ordinarily did. It was almost as if he and his wife had a kind of telepathy between them and responded to each other's movements and to each other's perceptions in a way far superior to the ordinary way. In the reality he was living at that moment there were no separations between things. All things flowed into each other.

Later that evening, at home, he and his wife sit listening to a Beethoven sonata. During many portions of the music he again organizes the universe in a different way than he does in his ordinary, everyday work. He organizes it in such a way that it is not he who is listening to the music; the music and he again are one. The music is inside as much as it is outside. He is not talking about it, he is not thinking about it, he is just intensely *being* with the music.

That night he goes to sleep, and during his sleep he has a dream. In the dream strange things happen. A kangaroo appears, hops around a mountain. Somehow it has the face of his older brother. He talks to it. The scene shifts and is now underwater. A beautiful mermaid appears. During the dream he does not question the "strange" things that happen. He knows they are right. He has organized reality again in a different way, a way in which all things are possible, all connections can be made. The symbol and the thing it symbolizes interact with each other constantly. This again is another state of consciousness, another reality in which our subject lives.

One of the fascinating things about alternate realities is that at the time you are really using one it makes perfect sense to you, and you know it is the only correct way to view reality. It is only common sense.

To use a modern phrase, the businessman was in an altered state of consciousness in these different incidents. An altered state of consciousness and an alternate reality are two different sides of the same coin: When I describe its rules and its Basic Limiting Principles (to use a term of the 20th-century philosopher C. D. Broad describing the basic assumptions of a view of reality), I am talking about an alternate reality; when I am perceiving and reacting according to these rules, I am in an altered state of consciousness. Each of us, during every day, uses several different constructions of the universe. We are in "altered states of consciousness," we are using "different constructions of reality," we are using "different metaphysical systems," we are in "alternate realities." All the evidence we have is that this shifting is essential to us. Certainly it is universal, it occurs in every culture and in every age we know of. If we encourage the use of alternate realities, as in meditation, play, serious music, and so forth, we increase the ability of human beings to reach toward new potentials. If we prevent it, we damage these people. This has been shown, for example, in the experimental work on preventing people from dreaming while allowing them to have the normal amount of sleep. The research had to be discontinued because it psychologically damaged the individuals.

The degree to which psychology, and the social sciences generally, have accepted the idea of different realities is very limited. Today psychologists generally do not affirm the idea of equal validity of these realities. The social sciences generally regard the sensory reality, the ordinary everyday state of consciousness, as the "correct" one, and the others as being due to some aberration or other. Literally they are "altered" from the "correct" one.[15]

In psychology such pejorative terms as "concretistic," "regressive," or "schizophrenic" thinking are used to describe the various "altered" states of consciousness, with the implication that the psychologist could probably cure them if he wished. In the background of this is Freud's dictum, "Where there was id there shall be ego."[16] Today a group of psychologists is even attempting to bring dreams under the active control of consciousness, working with what they call "lucid dreaming" (dream states in which you are aware that you are dreaming).

To the anthropologist it is clear that the native is indulging in "primitive" or "magical" thinking and that at any point where the anthropologist's and the native's thoughts about a particular problem differ, it is the native who has lost touch with reality. When sociologists discuss the difference among the "lower," "middle," and "upper" class orientations, they are generally rather clear that the middle-class views are the most effective and closest to the correct view, to "real" reality. The fact that sociologists are usually of middle-class origin is probably relevant here.

To the degree that they accept the fact that human beings live in different realities, the social sciences have adapted a procedure for investigating what these realities are. The procedure is to obtain as full a description as possible of the basic assumptions—the Basic Limiting Principles—of the particular reality. This is done in two ways. The first is to ask about them directly. For example, asking your informant, "Does 'Mana' obey the human will?" The other way involves observing and listening and figuring out what Basic Limiting Principles can be operating if the actions and words make sense.[17]

For example, if our businessman had prayed, "Please don't let it be meningitis," he was construing reality in such a way that prayers make sense and can be answered. The universe can respond to emotion if properly expressed. This then is one of the rules of this particular alternate reality. A third way, lately more popular, is for social scientists themselves to experience deliberately altered states of consciousness, as by the use of LSD or meditation, and then describe the cosmos as they perceive it during the height of their experience.

Social scientists, then, observe the way their subjects organize and perceive reality in various situations. They examine and try to analyze the structure and nature of the different particular organizations. They sometimes go on to define in what situations variations from "the correct view" (the social scientist's view) take place, as in the dream, play, psychosis, drug condition. In addition social scientists must make a determination between "normal" and "pathological" states of consciousness. Very little work has been done in this crucial area. It is plain, for example, that you if have 437 schizophrenics in a mental hospital, this does not mean that you have 437 different valid constructions of reality. It means simply that you have 437 schizophrenics. But how many valid constructions of reality are there? We personally believe, for example, that the number is comparatively small, but we know only a few of the rules for determining the validity of a construction of reality. The few rules we know are: (1) it must help you attain the goals recognized in the state as valid or answer the questions defined by its rules as real questions; (2) it must be internally consistent; (3) it must be—for we are human beings—a state of consciousness that human beings can use and in which, if only briefly, they can continue to function and survive. *Beyond that we do not know.*[18]

Social scientists are not concerned with one realm of reality or another. They are concerned with how subjects organize their total experience at a particular time in a particular situation.

Physicists approach the matter of alternate realities quite differently and even more diffidently. They divide the world into different "domains" for study. They identify these domains by such names as "mechanics," "thermodynamics," "chemistry," "plane geometry," "neurology," "psychology," and "sociology." In each domain of experience that they study physicists ask certain questions: "What are the observables in this domain?" "What kind of measurements can be made here?" "What are the laws relating the observables in this domain?" In the following pages we shall be discussing these matters from a more general viewpoint. We treat them in much more detail—and with more "scientific rigor"—in chapters 3, 4, 5, 6, and 7.

In each domain the entities, their observable properties, and their laws are different. All are compatible, none are contradictory, but they are different. Moreover, as certain boundaries between groups of domains ("realms") are passed, observables and the laws relating them become very different indeed—so different that in order to deal with them a physicist must use what is strictly speaking a different construction of reality. In our later chapters on reductionism this fact will be carefully analyzed. The physicist's data in certain realms can only be "explained"—made lawful—by the assumption that in these realms the universe must be understood according to a very particular organization of reality.

As a basic example of what we are talking about, let us consider the question: What happens when we proceed from the realm of things we can see and touch to the realm in which the entities involved are too small to be observed or touched, directly or with instruments? Since visual and touch characteristics are no longer present, concepts relating to them are no longer meaningful.

We take a blue billiard ball and shrink it to one-thousandth of its original size. It is now like a mote in a sunbeam, to use Vedic language. We shrink it another million times. It is now completely out of the see-touch realm.

Color is caused by the reflection of a particular wavelength of light in that range to which our eyes are sensitive. Our shrunken billiard ball is now smaller than these wavelengths. It cannot reflect light. What is its color? It has none. It does not even have an absence of color. The term simply does not apply, any more than the term "loudness" applies to a cloud or "weight" to a length of three meters.

"Texture," too, loses its meaning. How would you find out if the "surface" of our micro-billiard ball is now "rough" or "smooth"? As a matter of fact we are no longer even sure what we mean by this question, since we are no longer sure if it has a surface. How could one tell? You can no longer see or touch its surface even theoretically. And if you can tell nothing about its surface, then what is its "shape"? We can say anything we want about its "shape," but there is simply no possible way of finding out if we are "right" or "wrong." (If you say that the billiard ball is now "pie-shaped" and I say that it is "ribbon-shaped," there is no way of determining which of us is right.) The term "shape" has lost its meaning and relevance; it no longer applies. "Shape" is also a see-touch realm characteristic and must be considered as inapplicable in this realm of miniature onta.

Similarly the concept of "size" becomes a difficult and chancy one at this level. It is no longer quite clear what we mean by it. What do we mean by the question, "How large is an electron?" Are we dealing with a question similar to "How thick is the Equator?" (We do knows its length.) Or is this a situation in which the answer to our question is partially determined by the structure of the experiment, as the size of a comet is partially determined by its closeness to the sun (although its mass remains the same), or as the size of a blown-up balloon is partially determined by the surrounding air pressure. In any case size is not the essentially simple and stable characteristic that it is on levels where we can visually observe things. This is partly because "surface" is a visual concept and has no place here. We cannot describe the surface of a subatomic particle—we cannot even say where the surface is located (How far is the surface from the center of an *on* if we cannot determine its shape?) "Motion" also becomes something quite different in these nonvisual realms than it is in situations in which our eyes can serve us. Since the best we can hope to do is to find a sign that a particular *on* which was then in one place is now in another, we cannot say what, if anything, it did in between. Oftentimes we cannot even be certain it was the same *on*. There are exceptions to this: A cloud chamber track is always, and without conceptual difficulty, assigned to the same particle.

Since, as far as is known, a given subatomic particle may have no internal organization and since we cannot see its surface, there is no way of distinguishing one particle from another. As to its location, about the best we can do is to say in the physicist Arthur Eddington's terms, "it is smeared all over a probability distribution."[19]

(There is another unique feature that characterizes the members of the atomic and nuclear microcosm: Their *essential* properties are precisely the same. All electrons, protons, neutrons—indeed, all atomic nuclei—have, in accordance with experiments, the same mass and the same charge as other electrons, protons, or neutrons. The probable error in their values is extremely small. This is another reason for their indistinguishability. In the macrocosm such uniqueness is never encountered; no two bodies occurring in nature have precisely the same mass, size, or shape—unless they have been made by man. This fact, taken for granted by physicists, has never elicited an explanation from philosophers.)

We often say that an atomic electron is in a certain orbit, but this does not imply the visual characteristic of traveling from point to point in traversing the orbit. Nor does it "jump" from one orbital state to another in the sense of being between them while jumping.[20] (We will discuss this further in chapter 7.)

In realms of experience in which the entities are too small to be, even theoretically, seen or touched, concepts such as size, shape, surface, and motion change or even lose the meanings they had in the see-touch realm. "Locatibility" on these levels therefore also takes on new meaning. An *on* can be located when it has just interacted with an entity that is large enough to be visually perceived. An electron "hits" a scintillation screen, and we see the flash. We can say that the electron was interacting with the screen there then. That is in principle all we can say. Or, in a cloud chamber, a line of water droplets is formed. We do not see the electron; we see a large instrument that was affected in a particular way by an electron a short time ago. Where is the electron now? Again, "Its location is smeared all over a probability distribution."

We cannot define size, shape, identity, or locatibility in the domain of the very small (the "microcosm") in the same sense as we can with things that we can see. It is therefore reasonable to expect that the way things interact will also be different.

If we watch the interaction of two billiard balls on a table, we understand the cause-effect nature that propels one in one direction, the other elsewhere after they collide. We know which ball is which. If we had enough knowledge and were mathematicians, we could predict exactly their courses and the directions they would take and how far they would travel before stopping. As a matter of fact, that is exactly what a good billiards player does with astounding accuracy.

In the realm of onta, however, which we cannot observe steadily, but where we see effects of their presence only occasionally, where we often have no way of knowing which electron was which after two collide, where the more accurately we can measure the position of an *on*, the less precise we are about its momentum and vice versa, where visual characteristics do not apply, we would expect their modes of affecting each other to be different from the modes of billiard balls.[21] Since we cannot differentiate one *on* from another, we cannot predict what each one will do. Since they operate under law and not under whim, we can predict statistically what they will do.

To make this clearer let us take the following example. Suppose I am an engineer and I have a large number of machines, all, so far as I can tell, stamped out in the same factory under identical conditions and from identical materials. I cannot tell them apart. I cannot tell which one of them will break down first nor what part of it will first go out of order. However, if I have studied this type of machine for a long time and have all the shop records available from past shipments of them, I can predict with great accuracy how many of them will go out of order in any time period and in what ways they will break down. The basis of prediction for these imaginary machines has changed from cause-effect to statistical. The same change takes place when we go from realms in which we can distinguish things by eye to realms in which the entities are too small to be seen.

What have we been saying here? There is a way to describe how reality works that makes plain common sense when applied to the see-touch realm. This includes the characteristics things have, such as size, shape, and color, and how things happen—how entities move and interact. In this system how things are now absolutely determines exactly how they will be later, and if we have enough knowledge, we can make completely accurate predictions. When dealing with the sensory realm, this is the "correct" system to use, it is the "right" metaphysical system, the "true" description of reality.

When we are dealing with the microcosm, this system no longer applies. Entities have different characteristics, they move and interact in quite different ways. In this realm we must use a different description of reality in order to deal scientifically with the data. In the microcosm the new "metaphysical system" is the "correct" one—it is the "true" description of reality. (More will be said about this in chapters 8 and 9.)

Which system is *really* the correct one? It depends on the realm dealt with. The assumption that there is one "true" definition of *all* reality is outworn. As we shall show later, there is no contradiction between different valid systems of explanation—different valid realities. But they *are* profoundly different.

Because of the kind of measurements that can be made in each realm, the type of data that emerges, and the laws relating observables that must be introduced to make lawful sense of the data, the physicist finds that he must use either three or five different "realities" ("metaphysical systems," "alternate constructions of reality") to explain the data. We say "three or five" because he actually uses three, but if he extended his method further, he would need at least five. These are:

1. The see-touch realm—to the limits of instrumentation. This might also be called the "sensory" realm, or "middle range."

2. Things too small to be seen or touched even theoretically—the microcosm.

3. Things too big or whizzing by too fast to be seen or touched, even theoretically—the macrocosm.

These are the three realms to which physicists apply their method. There are at least two others to which it might be applied:

4. Meaningful units of behavior of things that are alive; that is, units of behavior above the reflex level. This is the realm of domains in which the organism seeks for food, runs to escape danger, mates, etc. (We define this realm and number five more carefully in chapter 12.)

5. Man's, including the physicist's own, inner experience.

In each of these five realms, if the physicist asks his questions ("What kind of measurements can we make in this realm? What are the observables here? What laws relating these observables to each other can we postulate?"), he finds quite different answers. As already indicated, in the see-touch realm a physicist can make quantifiable measurements, see clear causes and effects, observe that the present condition—the "state of the system"—inexorably leads to the next event, and use mechanical models. In the microcosm he can make quantifiable measurements but cannot observe cause and effect in the customary sense—i.e., in terms of predicting events—and cannot use mechanical models. If he extended his method to the domain of inner experience, he would find that he cannot make quantifiable measurements, can only observe cause and effect in the past but not predict specific events in the future, must include "purpose" as an observable, and cannot use mechanical models.

Each of these five realms has different answers to a physicist's questions: "What are the observables here?" "What are the laws relating these observables to each other?" In some of them, for example, time and space are different. Thus we cannot assume that space between the onta or between the galaxies is necessarily Euclidean.[22] How can we measure space? Personal space—the space that my inner life uses and that influences my behavior—is a far cry from "yardstick" space. A lover or a beautiful butterfly thirty feet away from me is "far." An uncaged tiger at a hundred yards is "very near." (If I know that the tiger is hungry, it is even closer!) Also in the realm of the very large or the very fast, time, space, size, velocity, and mass take on relationships entirely different from those in the realm of things that can be seen and touched.[23]

FIVE REALMS OF EXPERIENCE

Operation \ Realm	Very small	See-touch	Very big or fast	Meaningful units of behavior	Inner life
Quantification	yes	yes	yes	yes	no
Events called cause and effect (specific predictability)	no	yes	yes	no	no
Mechanical model applicable (entities can be visualized)	no	yes	no	no	no

At present physics is struggling with this problem. Much of the present speculation about "locatibility" (similar to the old term "action at a distance"), "hidden variables," the nature of onta (particles vs. fields), and like problems stems from the fact that in the microcosm and in the see-touch realm different constructions of reality are necessary. What is an insuperable problem in one of them—an impossibility, a miracle if it happened—presents no difficulty at all in another. Again, there is no one rationality, made by one rational God, which governs the entire universe.

It is immensely difficult for most of us to accept the fact that there is more than one valid way in which the world works. We are deeply conditioned to assume that we *know* the one truth and that everything else is, somehow, less real. To question this seems to us to be abandoning all reason and placing ourselves in a chaotic and unpredictable cosmos. It brings us to that "catastrophic anxiety" the psychiatrist Kurt Goldstein has described as the most severe of all anxieties. Goldstein has demonstrated how we, in our development, build our ego to support, and be supported by, that view of reality our culture believes to be the only correct one. If this model is no longer supported by the culture around us, or if we are faced with data that contradict it, we feel as if we are in great danger, the danger that our unsupported ego will crumble and come apart. We feel that we can no longer remain whole. From another viewpoint cultural anthropology has described the same phenomenon. Instead of "catastrophic anxiety," this domain includes the notion of "marginal man"; people who have actually and physically moved from the culture in which they were raised to a very different one, or have survived the destruction of their culture and now live in another. When individuals in this type of situation have been studied, it has been plain that they were subjected to very strong ego-destructive forces with consequent ego impairment and weakening of the ability to function.

Joseph Conrad, describing the idea that each culture shapes reality for the individuals born into it, and that they can survive unscathed only by staying within this definition of reality, wrote in his novel *Lord Jim*: "A man that is born falls into a dream like a man who falls into the sea. If he tries to climb out into the air as inexperienced people endeavor to do, he drowns . . ."

This shaping of reality is a process that profoundly affects the shaper. For example, included in our construction of reality is a definition of "human being" and the relation of this entity to the rest of the cosmos. The definition is applied to the person who makes or accepts it with dramatic results. Also included in the definition are answers to such questions as "Is the universe friendly?" and "Is it run by law?" They define the cosmos of each individual and therefore the way that he shapes himself to fit it. Omar asked, "Who is the potter pray, and who is the pot?" More recently Jung observed that it was not only Goethe who created Faust, but also Faust who created Goethe.[24] We are dealing with a circular and cybernetic system. When it is distorted and shaper and universe no longer fit each other, then overwhelming "catastrophic anxiety" results. The individual becomes profoundly disturbed.

We feel this sort of threat implicit in the idea that there is more than one "real" reality. Nevertheless, very often, if we scientifically follow the data and their implications, our older theories must be abandoned. In describing one aspect of this Werner Heisenberg wrote: "A consistent pursuit of classical physics forces a transformation in the very heart of that physics."[25] Science today, both in physics and in the social sciences, has brought us to the point where we must face the fact that if we wish to proceed on the scientific path and attempt to make our data lawful, we simply cannot have only one set of principles about how reality works. We need to allow for a number of alternate realities.

An understanding has gradually developed among social scientists in the past ten to fifteen years that we can and must use different reality organizations to deal with data. We began to understand that the organization we were using—that of Western "common sense" and 19th-century physics—was not adequate. As we began to comprehend that physics was using several different organizations of reality to deal with different kinds of data, a number of social scientists (including one of us—L.L.) made the error of believing that physics had solved our problem for us. We believed that we could use one of the alternate constructions of reality devised by physicists to fit *their* data to make *our* data lawful. Because these constructions—like those of quantum mechanics or relativity theory—were so different from the everyday construction we had tried so hard to use—and with which we had failed so consistently and because they did not lead to the same problems, we thought that one of them must be the system we had been looking for.

This was a mistake. Physics *had* shown that it is legitimate—and sometimes necessary—to use alternate interpretations of reality in dealing with different realms of experience. It had *not* devised a system that could be used to deal with meaningful behavior or with inner experience. Here, constructions of reality are needed that make the data from *these* realms lawful, not constructions borrowed from other realms. This mistake which some of us made was understandable, but it is now time that we leave it behind us. Neither the metaphysical systems used in quantum mechanics, nor that used in relativity theory fits the data of inner experience or those of meaningful behavior.

It is true that all the rationalities that are useful to science are compatible—none are contradictory even though they are very different. This however, must not obscure the fact that they are very different and the compatibility is probably due to the circumstance that it is human beings who are making these constructions of reality and who provide the coherent consistency that runs through all of them. Einstein wrote that the greatest miracle of the universe is its comprehensibility. The reason it is comprehensible is that we can only know it as it is construed by human beings and our own works are comprehensible to us. (This will be set forth in detail in chapters 3–6.)

From the viewpoint of present day science, then, to sum up, there are two basic ways of organizing the knowledge of alternate realities. The social scientist works from the viewpoint: "What metaphysical system (state of consciousness) does this person impose on reality? Under what conditions does this system change to another? What is the other? Which systems are normal and which are pathological?" The physicist asks: "What metaphysical system must I use in this particular realm of experience? How can I find the laws which make different systems compatible?"

In the social sciences, there is the examination of the general set of rules by means of which an individual organizes his total experience. It is found that these differ at different times and in different situations. In physics there is the examination of the data in a particular realm of experience followed by the devising of a construction of reality which will make these data lawful. It is found that this construction must differ considerably in certain different realms of study.

The two approaches to the problem must ultimately meet and synthesize in the study of inner experience. In this realm, the method of the physicist, asking what kind of measurements can be made here, what are the observables found, what are the laws relating these observables to each other, already partially meets the method of the social scientist asking what laws and construction of reality is the individual using: How is he construing reality? When the two approaches coalesce in the realm of inner experience, we will have the beginning of a real science of psychology. In this book we endeavor to approach this goal.

2 Structures of Reality: Domains and Realms

> Order, unity and continuity are human inventions, just as truly as catalogues and encyclopedias.
> —BERTRAND RUSSELL
> We must avoid here two complementary errors: on the one hand that the world has a unique, intrinsic, pre-existing structure awaiting our grasp; and on the other hand that the world is in utter chaos. The first error is that of the student who marvelled at how the astronomers could find out the true names of the distant constellations. The second error is that of the Lewis Carroll's Walrus who grouped shoes with ships and sealing wax, and cabbages with kings . . .[1]

IN HIS *Micromegas* Voltaire has an immensely wise alien come to visit earth. He has dozens of senses and can thus perceive much more of reality than humans can. He comes from a race devoted to the acquisition of wisdom and that lives thousands of years. On leaving earth, he leaves philosophers there a book containing all the knowledge that can ever be gained about the pure and ultimate nature of things. It has only blank pages.

From the viewpoint of modern science there is no such thing as an ultimate and unchangeably correct description of reality any more than there is any such thing as an ultimate and correct shape of a lump of clay. The question "What is the true shape of the clay?" is meaningless. Its shape is whatever it is shaped into. One shape is as valid as the next. To some extent this is true of reality.

Half a century ago scientists used to evaluate their theories and theoretical concepts by asking the question: Is the theory or theoretical concept, true or false? Since the days of Henri Poincaré, the notion has gradually gained ascendency that the aforementioned criterion is not the proper one. Nowadays we do not ask whether a given concept is true or false. We ask: Is it convenient or inconvenient; is it useful or not.[2]

There are clearly limits to this definition of reality. We cannot shape a lump of clay into a self-supporting arch a yard high and a twentieth of an inch thick, or into a flower with thick petals and a long, thin stem. The clay will not hold these forms. It will break. Reality also has its limits. If you organize it in invalid ways, you are no longer dealing with reality, but with schizophrenic projections. Indeed, it may well be that all we will ever be able to learn about the nature of the clay of reality will come from determining the limits of how we can organize it. We can, perhaps, find the outer boundary of its possibilities, but we can never determine the "true" shape and order of reality. We shall have to give up this dream.

There is "a game of give and take between consciousness and outside reality,"[3] and we can only know that anything and everything we perceive and react to is a synthesis of the two. We cannot see either individually any more than we can study a culture without individuals, or individuals without a culture.

The pragmatically oriented philosopher will say, "The true and valid shape of the clay is the shape that works best." This seems to us a reasonable statement, but the question arises, "Works best for what?" A cup, a saucer, an insulator, Rodin's statue *The Kiss* may have very different shapes, each working best for its particular purpose. Similarly, different methods of organizing reality will work best for different purposes.

The older view of consciousness is that it is something that arrived late on the scene of history and found everything already well organized. Consciousness then *discovered* reality. From the viewpoint of 19th-century science, consciousness was something with bad hearing and defective eyesight trying to figure out how inorganic nature had put the world together.

Today, science is beginning to view the nature of consciousness quite differently. Classifying and organizing the world are seen as human activities. What we can observe of reality is our own organization of it. Reality is a compound like water, with consciousness one of the elements. But we can never hope to know what the compound would be without consciousness. If we were to say, "We now have an idea as to what reality is before consciousness was added. Let us check and see if our idea is correct," the second sentence would have no meaning. We cannot even conceive of a way to make this check. There is simply no way to test the accuracy of our guess—[4] although we can prove it wrong, or more likely than some other guesses, we can never prove it correct.

No longer do we view consciousness as a late arrival on the scene. Consciousness *is* the scene. It is no longer something that coolly and with detachment (or hotly and with passion, for that matter) observes reality. It first organizes reality, and whatever the word "real" may mean, this organization is real.

The reason why our sentient percipient, and thinking ego is met nowhere within the scientific world picture can easily be indicated in seven words: Because it is itself that world picture. It is identical with the whole and cannot be contained in it as part of it.[5]

One writer described an aspect of this evolution of science in this direction and to his viewpoint as follows:

Put the matter another way: If man lived under a tinted plastic bubble and had no inkling of anything outside it, he would have no reason to doubt that the sky had, in itself, the tint his bubble imparted to it. The world view of primitive man was such a bubble. Never encountering men who lived in cultural worlds that were radically different from his own, he assumed that the meaning that pervaded his world was objective. He had not invented it; it came with life, built into the ribs of being as the skeleton on which all else hung.

We no longer see things this way. Traveled in time and space, we have come to know and covenant with men whose plastic domes are tinted otherwise than our own. Thus enlightened, we can no longer assume that the sky is, in itself, the color it appears to us. The color we see is not a mirror-image of what characterizes the world apart from us. Instead, what we see is the convergence of *something* in the world with our socio-psychophysical brand of sensing apparatus. It is in part a construct.[6]

We tend, of course, to believe that what we see is what is real and true, is actually "out there." This belief was termed by the philosopher Nietzsche "the fallacy of the immaculate perception." Scientifically, however, we know that there is no such thing as an innocent eye—our eyes create much of the form of reality we perceive.[7] (This field of inquiry, called epistemology, will be treated more fully in chapters 3–6.) Bertrand Russell wrote in this connection:

The supposition of common sense and naive realism that we see the actual physical object is very hard to reconcile with the scientific view that our perception occurs somewhat later than the emission of light by the object; and this difficulty is not overcome by the fact that the time involved, like the notorious baby, is a very small one.[8]

"Common sense," wrote the philosopher Susanne Langer, "is the popular metaphysic of one's generation.[9] Einstein called it "that collection of prejudices accumulated by age eighteen."

Nevertheless, wherever we look there is organization and law. We and the rest of what is are organized and interact in definite and coherent ways. When we organize reality, we do not do it according to whim (unless we are so damaged that we belong in a mental hospital), we must do it according to law. Only in this way can we have the coherence and stability so necessary for human beings. Otherwise we cannot continue to exist and maintain our own organization; otherwise the cosmos is not viable for humans.

The universe without the illumination of consciousness would be, in Einstein's words, "a mere rubbish heap."[10] We invent and discover its continuities and discontinuities, its coherences and its "spots and jumps." It ("reality") is somehow there, but we alloy it into being with our consciousness. The two things that filled Kant with ever-increasing admiration and awe, the starry heavens above us and the moral law within us, are both our own invention-discovery. They are the creation of an order, of depth and beauty which show how artistically consciousness can affect the raw stuff of reality. The moral law, the starry heavens, the wheatfields at Arles, $E=MC^2$, *The Magic Flute*, and the Pythagorean Theorem are all examples of the aesthetic creativity we can bring to the construction of the cosmos in which we live.

Perhaps one reason why it is so difficult for us to accept the idea of multiple, equally valid realities is that so much that made the world stable and permanent, Gibraltarlike and reassuring, has already been taken away from us. The earth was unique, and the center of the universe until the work of Kepler and Copernicus told us otherwise. Man was descended from God and made in His image until Darwin told us otherwise. Our reason was an absolute tool with which we could understand the world and logically react until Freud told us otherwise. Our customs and social beliefs reflected eternal verities and told us what was absolutely right and what was absolutely wrong until the modern anthropologists told us otherwise. What was left was the idea of truth; that there was one truth and that, at least, was unique, stable and eternal. One poet wrote:

> It fortifies my soul to know
> That, though I perish, truth is so.

And now science seems to be taking this last bastion of stability in a frightening universe away from us. Is it a wonder that we resist? However, if we are going to use science to help us build a world fit for human beings—and human beings fit for the world—we must base it on a *human* world picture: a world in which a machinelike cosmos rules machines and a humanlike cosmos rules human beings.

We do not realize the extent to which we have been *taught* how to see the world. We have been taught how to *organize* our perceptions and relate them to each other. Even the see-touch realm, which is so "obvious" and "clear" to our perception that we are absolutely certain of the truth of what we see, is largely what we have been taught. We have learned from our families and our culture what is good and what is evil, what is beautiful and what is ugly. We have even learned what things look and sound like. Tell me what sound a rooster makes when it greets the morning sun and I will be able to make a pretty good guess as to what country you spent your childhood in. If you say "Cock-a-doodle-do," I will guess the United States. If you say "Cocorico," I will guess France; if "Kikerikee," Germany; if "Kukeriko," Israel. Or if you ask a Frenchman what is the sound of raindrops on a window pane, he will say, "Plouf, plouf." Ask a Japanese and he will reply, "Zaa, zaa." Ask the Frenchman what sound a contented cat makes and he will tell you, "Ron, ron." Cross the border to Germany, ask the same question, and the answer will be, "Schnurr, schnurr." When the cat is asking for its dinner, an American will hear it say, "miaoww," a Japanese will hear "niago." Further, a Japanese mother would be very surprised if her child's first words were "Ma ma." She knows a child's first words are, "Ogya, ogya."

Take your dog around the world and ask what sound it makes when it barks. The answers will vary with the country. In France it will be "Gnaf, gnaf"; in Spain, "Guau, guau"; in Japan, "Wung, wung." Which of these answers is correct? None and all. Isn't there a "right" sound that the dog makes? How would you find out what it is? Shall we get a judge from Africa? He will tell us that the dog goes, "Kpei, kpei."[11]

Much study in the present period has been devoted to the way our language affects, and in part determines, what we can perceive and how we organize our perceptions. It is interesting to speculate how different the world must have looked before Thomas Gray coined the word "picturesque" in 1740 or before Whewell coined "scientist" and "physicist" in the 19th century (or before Shakespeare coined the words "assassination," "disgraceful," or "lonely.")[12]

One sign of the change and development of our understanding in a particular area is that questions once considered to be important come to be considered invalid or uninteresting. The reverse is also true—questions that were considered trivial or meaningless become central and of great concern. The Greeks of the classical period did not consider the origin of the universe a problem. Before Descartes, no one felt that the relationship of the mind to the body was a valid problem. Darwin once characterized the problem of the origin of life as "rubbish."[13] There was an ongoing debate at the court of King Louis XIV as to whether two "perfect" painters, using the same model, would paint identical pictures. The Albigensian Heresy, for which hundreds of thousands were once killed, does not concern us today. Thomas Aquinas, surely one of the most powerful intellects known to human history, gave two-day seminars on such subjects as "Is God's knowledge the cause of things?" and "Do angels know the future?" Most of us no longer consider these questions to be interesting. Similarly, such problems as "What holds the Earth up?" and "Which is the top and which the bottom of the Earth, and why don't those on the bottom fall off?" are no longer meaningful. The march of our understanding has taken us past many questions. In like vein the question "What is reality?" "What is its true shape and nature?" is one for which we no longer expect a single answer.

The great discovery of the ancient world (about 600 B.C.) was that there was an intelligible structure to the world. The great discovery of the Renaissance in Europe was that we could use this structure for our purposes, that the more we understood it, the more we could control matter and energy. The great discovery of the present-day revolution is that—within limits—the structure is up to us, and different formulations of it must be used with different types of experience and to attain different goals. One way to characterize the viewpoint here suggested is to compare a conception of reality with a work of art. One cannot say whether a painting is correct or incorrect, one can only ask if it accomplishes its purpose; if it adds to and enriches experience.

To clarify this further: Suppose a group of painters—da Vinci, Vermeer, van Gogh, Titian, Magritte, and Toulouse-Lautrec—all painted the same model from the same viewpoint. So did Smith, who has training but little talent. It is hardly intelligent to question the correctness of each painting or to ask which better portrayed the model. The relevant question is, Which one would you rather have in your living room, bedroom, study, local museum, or trashcan. Santayana once wrote, "Critics quarrel with other critics. With an artist, no sane man quarrels." The analogy could be continued. All conceptions of reality are not valid, and all collections of form and color in one place are not works of art. Both must fulfill definite requirements. (Exactly what these are we are not sure in either case.) An old palette full of paint stains is not a painting—it is a mess. The difference is that you encounter a lot more trouble if you accept an invalid conception of reality as valid than if you accept the old palette as a work of art.

The scientist, like the artist, is constantly faced with the problem: "How do you organize and comprehend the world?" Although this is the central problem of both science and art, it is rarely accepted fully by either the scientist or the artist. Generally both function by accepting the standard cultural organization and comprehension of reality and, working as technicians, doing the best they can to analyze and improve it to the maximum, so that it can best help human beings achieve their goals. Only rarely in either field (and then generally in both) do people devise a new organization and comprehension of reality. And when this does occur, it is done slowly, with different rates of advance in different areas, and it has a profound effect on human activity. The change of comprehension and activity may be well underway before there is a clear statement or philosophical understanding of the new organization. Once a clear statement is made, however, it seems to extend the change and to make it easier to deal with by workers in various fields.

We can clearly see examples of this kind of change of perception of, and reaction to, the world in the 14th to 16th centuries in Europe and again in Europe and America at the end of the 19th century. In the Renaissance we observe the change from a Medieval world, perceived in terms of, and built around, small enclaves of highly traditionalized human activity, communities isolated in space by vast stretches of forest while connected by an overarching religious meaning and set of principles. Further analysis of reality, both artistic and scientific, dealt primarily with these religious aspects.[14] From this set of perceptions and reactions there developed in this period a new orientation focused on individual differences and activities, a primary direction of greater understanding and control of what is accessible to the senses, and a breakdown of the isolation of the small groups of human beings into much larger groupings with much more possibility of nontraditional, individual activity.

The profundity of this change can hardly be overestimated. At the end of two centuries individuals were perceiving and reacting to the new organization of reality so naturally and completely that only specialized scholars could understand the way people had lived previously, the problems they were concerned with, and why they behaved as they did. In one of his best stories, "The Thousand and Second Night," Edgar Allan Poe had Scheherazade tell the Sultan a new tale of wonder. In this one she did not talk—as she had in the previous tales—of Sinbad and Aladdin, of magic and sorcery. Instead, she told of the works of 19th-century science such as the telescope, the telegraph, and the steam engine. The Sultan responded that her Thousand and One Tales had been believable, but that this one was preposterous!

Toward the end of the 19th century another shift in the Western approach to reality appeared. In art, the Impressionists began to break down the surface appearance of things and to go beyond immediate sensory experience. From physics to psychology science began to do the same, shifting concentration from mechanics to atomic structure and from symptoms to psychodynamics. It was consistent that at the same time this search below the surface, led by the Expressionists, characterized art, and similar movements were in progress in various fields of science, the most popular figure of literary invention was a man who explored below the surface of situations and events—Sherlock Holmes. The change in orientation is as profound, and as irregular in its advance, as that from the Medieval to the Renaissance viewpoints. One difference is that the change brought about at the end of the 19th century is still going on and is very much a part of our lives today.

The major, general statement of the organization of knowledge of the Medieval world was the Aquinas-Aristotle synthesis that put together Greek science and Christian theology in one coherent statement. The Renaissance viewpoint was defined by Descartes' formulation of its basic dualism and Comte's clarification of half of this dualism into specific sciences. These two show us how the Renaissance viewpoint organized *knowledge:* how this worldview perceived and reacted to reality. It separated objective and subjective perceptions and divided the objective into such specialties as physics, chemistry, biology, and sociology. Although, as Comte showed, each area of specialization seemed to rest on the one below it in the so-called hierarchy, they were seen essentially as separate. This naturally led to separate departments in educational institutions, intensive specializations in study and research, and a gradual separation of classes of specialists, with the communication between them growing more and more limited and confused.[15]

This separation of reality into objective and subjective knowledge, and the subdivisions in the objective sphere, led to tremendous advances in knowledge, prediction, and control of the "objective" world. So much progress was made in the ability to transform and distribute energy and to devise new forms of material and machines that what would have been regarded as sorcery in 1600 was commonplace 300 years later. From physics to medicine more progress was made in 300 years than in the previous 3,000.

Slowly problems of, and inherent in, the Renaissance breakdown of reality began to appear. Knowledge, prediction, and control on the "objective" side of the Cartesian dualism were proceeding at a very rapid rate and on the "subjective" side at a very slow rate. The division of reality into a sphere of matter—the *res extensa*—and a sphere of mind—the *res cogitans*—provided a very powerful methodology for the study of one and a very inadequate methodology for the study of the other. As we can see now but could not have predicted then, this imbalance had inexorable and unfortunate consequences. Our power to manipulate and control the "outside" world—matter and energy—advanced greatly, but we made no advances in the understanding of our own behavior and our inner experience. If a system makes a basic description of the cosmos by stating, in effect, "This is me (my consciousness) in here, and the rest of reality (matter) out there," as did the Renaissance approach, it can only analyze the "out there," not the "in here."

Each system of organizing knowledge develops partially in terms of, and in response to, a specific set of problems. These could not be solved by the previous organization and are usually the most pressing and vital problems of the culture. Thus the Renaissance organization developed in part because it became absolutely necessary to solve the problems of learning to deal effectively with the environment after the Black Death had killed more than one fourth of the population of Europe and the Medieval perception of the world was completely ineffective in doing anything about it. Every technique of the previous organization of knowledge (the Medieval world view) was used in the attempts to control the plagues—prayer, ecstatic mysticism, scapegoating, medicine based on sympathetic magic, and so forth. All failed. Under the pressure of a critical problem that it cannot solve with its present conception and organization of reality, a culture either develops a new one or, as the historian Arnold Toynbee has shown, goes under. Under the pressure of its critical problems Western Europe developed a new way of organizing reality, a way that made it possible to study and control the outside environment it defined—the *res extensa*, the world of matter and energy. This way made it possible to solve the critical problems of the period. Gradually, however, the problems inherent in the new organization of reality became crucial. Vast increases in knowledge of medicine and physics made the population increase very rapidly and also greatly developed our power to destroy ourselves and the balances with nature. The lack of advances in the understanding of human behavior and inner experience prevented any development of our ability to control the population explosion, to stop the poisoning of our planet, or to prevent wars. As one of us has shown elsewhere, there has been no advance in the understanding of the causes of war since Hellenistic times.[16]

As the Renaissance conception led to ever greater advances in medicine and the control of matter and energy, the problems in human behavior became more and more crucial. We are now very much in the position of the Medieval world in the 13th and 14th centuries. Our most pressing problems cannot be solved with the old world picture. We must either find a new way to organize knowledge, a new way of conceptualizing reality, or else go under. The Renaissance approach, so powerful and successful in other areas, has proved unfruitful, since advances in our problem areas depend on the understanding and control of the subjective side of the dualism, of human behavior and experience.

The new organization of knowledge is available to us. Its advantages include the fact that it has a place for consciousness and the study of inner experience inside itself. It also provides an effective method for the study of meaningful behavior. But it must be said again that any new model is terribly hard to accept as valid. We are so tied to the world picture we were taught as children that any suggestion of another—or, as in the present case, that the one we grew up with is valid for only one part of reality and that others are needed for other parts—is automatically judged to be nonsensical. After 2,000 years during which it was believed the most basic common sense that Euclid's geometry was *the* valid geometry, mathematicians such as Bolyai and Lobachevsky presented systems of geometry different from Euclid's, in the hope that these systems were worth considering in connection with the actual properties of space. They attempted to show that Euclid's system was valid for only a small part of reality and that other systems were needed for other parts. At first they were regarded by other mathematicians as neither serious nor sane.[17] Our first instinct is to reject the new, particularly if it implies a number of world pictures, and to say with complete conviction, "Our commonsense model is the true description of reality, and even if it were not, there is only one truth even if we do not know what it is. A thing is either true or it is not, and truths are universally applicable." This credo, now rejected by modern physics, still carries a deep ring of truth for us.

In the organization of knowledge we are presenting here the cosmos is divided into *domains* of experience. In each of these certain observables "appear." Some domains bear a sequential relationship to each other, and when this is true, a number of definite statements can be made about their relationships. Domains fall into larger groupings called *realms*, and each realm has a special organization of reality (metaphysical system) which is necessary to make the data from it lawful. We shall illustrate each of these points in some detail.

In the domain of "mechanics" we are dealing with a limited number of physical entities. Among the observables that appear when we work in this domain of experience are "mass," "position," and "velocity." We do not ask where these observables come from; they "appear" in this domain.

If we are dealing with a much larger number of interacting entities such as molecules, we enter a domain called "thermodynamics." A new set of observables appears. Here we have "pressure," "temperature," "free energy," "entropy." Again we do not ask where these observables come from; they appear.

In a simpler example, the concept of triangularity has no meaning when we are dealing with a domain with only two bodies in it. When we have three, triangularity becomes a new observable based on the relationships of the three bodies.

When domains are related to each other on the scales of size or complexity, they are often said to form a hierarchy. Under this condition observables in one domain are usually not conceivable or predictable from another domain. (The general relation between domains or realms is set forth in detail in chapters 7, 8, and 10.)

Looked at in the other direction, however, we can see how the observables in the second domain are explicable in terms of the observables of the first.

The existence of temperature, entropy, pressure, and similar observables in the domain of thermodynamics is not predictable from the domain of mechanics. Once we know of their existence, however, we can look to mechanics and see how the measurements of each of these are explicable in terms of the observables in mechanics (i.e., in terms of mass, velocity, and position of the individual molecules.) We cannot predict the existence of "triangularity" from the domain of only two bodies. Once we have a domain of three bodies and triangularity has "appeared," however, we can see how its measurements are explicable in terms of the observables of the two-body domain (i.e., direction and distance between the bodies.)

A melody is an observable that comes into existence when a certain set of successive notes is combined in a certain way. We can see how the melody is dependent on the notes and their succession. Study on the level of single notes, however, could not predict the existence of melody. Also, if we know the melody, we cannot predict the specific notes that compose it. It could be in various keys. But if we know the notes, we can determine the melody. Similarly, if we know the amount of each sale made in a store for a week, we can figure out the total volume of sales for that week. The reverse is not true: we cannot find out how many sales there were at what prices from knowing the total sales volume.

When we are dealing with one person, the observable of group behavior (group dynamics, etc.) is nonexistent. It simply does not exist in this domain, and its existence could not be predicted from it. We can observe certain things about the psychology and behavior of one individual or of a number of isolated individuals. Once the individuals become related, however, an entirely new set of observables (the phenomena of group behavior) springs into existence. The existence of these observables was not predictable from the study of the individuals concerned. We can say that a group behaves in a specific way because of the specific individuals in it and can demonstrate the truth of this. We cannot, however, predict from measuring the variables of group behavior (e.g., "group cohesion") how specific individuals in the group would behave.

To repeat, when the domains are related, it is possible (once the new observables appear) to predict measurements in one direction only. We have shown this in the examples of the melody and the sales volume. Further, as we have indicated, if we knew the position, mass, and velocity (the observables of mechanics) of every particle in a container of gas, we would also know the pressure, temperature, and other observables of thermodynamics. But the reverse is not true. If we know the pressure, temperature, entropy, and free energy, we cannot predict from these the position, mass, and velocity of each particle. There are simply too many possible combinations of these observables which could make up the particular measurement of the thermodynamics observables. The importance of this general relationship will become clear when we examine the relationship of consciousness and brain states.

One additional important general law about domains is: Observables appearing in any domain are lawfully related to each other. The relationships of the pressure, volume, and temperature of a gas exemplify this. This fact can give us some important tools for research. If, for example, group dynamics and parapsychological events appear in the same domain (more than one person and relationships between them), then we should search for the laws relating these observables and thus gain a valuable addition to our techniques for understanding both.

In the viewpoint of modern science no domain of experience is more "real" than any other. Each is exactly as valid as the next. "Nature has neither kernel or shell," wrote Goethe, and the present scientific viewpoint is in agreement. A domain is chosen in view of what one is trying to do.

In one domain Monet's *Waterlilies* is an assemblage of brushstrokes of oil paint on canvas. In another it is a magnificent painting. Neither domain is more real than the other. Counting the brushstrokes is a perfectly useful activity for certain purposes; so is enjoying the painting for others.

In one domain a blazing fire is red hot. In another it is the combination of a large number of molecules with oxygen. Neither "fire" is more real than the other.

Even when domains are related in a sequential order, one is just as real as the next. Since explanation is continuous from mechanics to thermodynamics, both are equally valid. Thermodynamics is not "nothing but" mechanics any more than violin playing is not "nothing but" the dragging of the hairs of a horse's tail over the guts of a cat. In the older science, as in much Eastern philosophy, some domains were considered more real than others. The "less-real" ones were seen as reducible to those "more real." These less-real ones were the equivalent of the world of illusion, the "veil of Maya" in Eastern thought, that hid the truth of the "ultimate" domain, the "true" state of being. There has been much confusion in science because of this now-outmoded doctrine. "Reductionism," the practice of attempting to account for all the properties of highly complex systems in terms of their simplest components, has been a part of our cultural thinking for several centuries, and has only lately been abandoned by science. (We discuss this way of thinking at length in chapters 7, 8, and 10.)

In the methodology developed here, new observables are approached in a consistent way, rather than being set off, apart from the rest of reality, making their realm inaccessible to study (as we have traditionally done with consciousness). Let us consider this new methodology in connection with our already familiar example. When we go from a domain of one-dimensional experience to a "two-dimensional domain," a new observable—"area"—appears. We do not ask where it came from. It was unpredictable from the first domain, present and inexorable in the second. We ask, rather, "What is the relationship of this new observable to others in the one-dimensional domain (direction and distance on a line)? What are its relationships to other observables that have appeared in the two-dimensional domain (such as angles and geometric forms)? What are its relationships with observables appearing on the three-dimensional domain, the next adjacent domain (such as volume)?" As we increase its "extensibility," its relationship with our other constructs, we learn more and more about it. This approach to new observables (as we shall see in chapters 12 and 17) applies to the observables of consciousness as well as to those of any other realm.

Let us look briefly at some applications of this viewpoint to the observables of our inner experience, of consciousness. In accord with what we have said about the laws relating sequentially organized domains, we can say that the existence of these observables could not have been predicted from the domain of the biological organism. Once their existence is known, we can determine the variables of consciousness from measurements of variables of biological functioning—but not the reverse. For example, we can look at certain measurements of the Central Nervous System ("brain states") and confidently predict that the individual is "depressed." We cannot, however, look at a depressed individual and predict with any confidence what the brain states will be: There are simply too many possible reasons for the "depression."[18] To carry this further, we can look at certain measurements of body chemistry and predict that an individual is "hallucinating" (as, for example if we find a high percentage of LSD in his blood), but we cannot go from the fact that an individual is hallucinating to the prediction of what the blood chemistry will be: There are too many possible reasons for hallucinations. Or we can examine the brain structure and sometimes predict confidently that the individual is feebleminded, but we cannot predict from the fact that an individual is feebleminded what his brain structure will be. We cannot say that all individuals who are depressed have a particular brain chemistry, or that all who are hallucinating have ingested LSD, or that all who are feebleminded have a particular brain structure. This relationship has important implications for research design and theory building.

One aspect of this is related to the fact that an individual's inner experience of life is organized as if it were a series of sequentially organized domains that might be termed "My childhood, high school years, the time I was drafted, my first job, when I married," etc. Here, as in the other sequentially organized domains we have been discussing, prediction is possible only in one direction. We can carefully examine any unit of inner experience (or of meaningful behavior, since there are lawful relationships between inner experience and meaningful behavior—see chapters 12 and 17) and show that it related lawfully to past experience and was inevitable. It was determined. (We can do this mainly through such psychoanalytic tools as free association.) We cannot, however, predict specific future experience or behavior with any confidence. If we have an adult who demonstrates a specific psychopathology, we can show that it was caused by a specific constellation of factors in his childhood. We cannot, however, look at a child with that constellation of factors in his home and predict that as an adult he will show that particular psychopathology. In a homosexual male we may be able to show that his sexual experience and behavior were related to an aggressive mother and a passive father. We cannot say, however, that all male children raised by passive fathers and aggressive mothers will become homosexual adults; many will become heterosexual. We can show that the experience and behavior of a specific adult criminal was directly related to the fact that he had a broken home and was raised in a slum ghetto. We cannot say that all children from broken homes in that ghetto will become criminals. A tremendous amount of research time has been spent in the social sciences trying to make predictive generalizations that, from the viewpoint we are presenting, are impossible. In the realms of inner experience and of meaningful behavior the past is determined, the future is not.

With any individual we can observe that adult experience and behavior are dependent on, and compatible with, childhood experience. However, from childhood experience we cannot predict adult experience.

To generalize further, so far as human experience and behavior go, the present can be seen as continuous with, and determined by, the past, but the future cannot be predicted as new observables appear whose existence is, in principle, not possible to predict. From the viewpoint of Domain Theory this is seen in the same way as what is often called the hierarchical organization of some domains of knowledge. Columbus discovered America, and the African slave trade resulted. Ford linked the assembly line to the production of automobiles and changed the sex habits of Americans. These sequences could not have been predicted at the time they were initiated, because new observables appeared as the sequences progressed. Looking backward, it can be seen that they were inevitable and determined.

As we indicated earlier, there are certain points, marking a discontinuity between domains, where a new system of construing reality *must be used in order to make the data lawful*. At these points we pass from one *realm* of experience to another. The point between the domain of the total biological individual and the domain of inner experience—consciousness—is such a discontinuity. We look later at some of the differences between the construction of reality needed for the realm of inner experience and the construction needed for the realm in which we deal with the biological individual—the see-touch realm (chapter 17). We also discuss there the realm of meaningful behavior and show why a separate construction of reality is needed in order to make the data from it lawful.

The Search for II
Scientific Truth

W E ATTEMPT IN THE BOOK a scientific synthesis of a variety of realms, a coordination of a number of important fields of knowledge including such nonphysical, nonmaterial subjects as sociology, economics, ethics, psychology, psychiatry, and even parapsychology. In these areas of investigation debate is common, different and often incompatible views are held, and answers to a given question are sometimes contradictory. No attempt to unify them is possible unless we understand a few of the basic elements that underlie *all* of science, all of verifiable knowledge. These elements are not commonly regarded as part of the various realms, the various sciences themselves, for there still exists a deplorable schism between philosophy and science. (Nor is there general agreement on other issues among philosophers themselves.)

To establish this foundation of understanding we first examine the sciences in which we find the most clearly and generally accepted methods of research—where procedures of validation and verification are definite, and where debate as to the so-called truth of results is minimal—and which share a common methodology. These are the physical sciences in the largest sense, including physics, chemistry, biology, and related subjects. We must acquaint ourselves with the way in which these sciences probe the validity of their beliefs before we proceed to our study of other realms.

Philosophers call the search on which we embark epistemology, the theory of knowledge. Their writings bristle with technical terms and present a variety of conflicting theses such as positivism, empiricism, idealism, existentialism, and many others—we cannot present them all. But one appears to us to be the only one supportable by modern evidence, especially by recent developments in physics and astronomy and particularly in quantum theory and relativity. It is the most neutral view, the one which, with an appropriate terminology, leads to only minor conflicts with the most prevalent forms of epistemology. For more on this methodology see *The Nature of Physical Reality* (Henry Margenau, 1979).

The terminology used in the next chapters may seem strange. Its use, however, is deliberate, for there is no field of learning where a single word (like "truth," "hypothesis," "fact," "postulate," even "definition") has as many different meanings as in philosophy. Nor are these meanings accepted as single-valued terms in any science. Consider, for example, the word "experience," the subject of chapter 3. If a person is not qualified for a job, we might use the perfectly meaningful phrase "no experience." In our sense of this word—and in that of most philosophers—this statement would mean that the applicant is dead.

Truth and fact are usually taken to mean the same thing, but in ordinary parlance we would have to say that the fact is that the fact and truth are different concepts. A mathematician, for example, will regard the statement that there is an infinite number of integers as true, but he would hardly call it a fact, certainly not in any literal sense of that word.

The meanings of the words "sensation" and "sense impression" are straightforward, but meanings are quite different from what is implied by "sensational." We therefore found it advisable in certain contexts to avoid the word "sensation" and to replace it by the technical term "protocol" or, occasionally, "perception." The word "protocol," together with its origin, will be explained in detail. As to the meaning of the verb "to perceive," we note here that it fails to distinguish between external, or sensory, awareness and awareness of conscious states. Our language permits us to say, "I perceive a cloud in the sky," but also, "I perceive that you are happy." The two senses of the word are clearly different.

A few further examples will be elaborated in this section: The word "temperature" may mean a feeling in the skin or the reading of a thermometer scale; time may refer to a subjectively experienced interval of waiting or to something visually observed on a clock; distance can denote a stretch of landscape surveyed by the eye or a line measured with a yardstick.

Even when I say, "I see a tree," some philosophers may correct me and claim that I *see* only certain shapes and colors while, if I am touching the tree, my hand has a feeling of solidity. The tree itself is the "essential" carrier of the impressions that assail me. It is the postulated object that has the seen qualities.

We now turn to more specific questions: When is a theory or an explanation correct? What is meant by "cause and effect"? Are they things, events, or mere theoretical notions? What makes a theory correct? Is a theory a set of facts, of observations, of perceptions, or is it a group of assumptions and related ideas? Too, verification is an important ingredient of every science, especially the realm of science here under consideration. What precisely does it involve? And finally, what is a definition? Does it designate a finite (or even infinite) set of objects, each having specified properties? Although some logicians propose this meaning, accepting it would make it impossible for me to define my feeling or my mood. The very meaning of existence is different in the sciences of mathematics and physics, even in physics and common sense.

Our teaching "experience" has led us to believe that a discussion of such abstract matters can be facilitated and made vivid by the use of diagrams. One feature occurs in all our diagrams, a vertical line called the "P-plane." The line is meant to be the projection of a plane extending upward and downward at a right angle to the diagram. To the left of it lies the "C-field," which contains the external world and all concepts it involves or generates. The right side of the P-plane is left vacant, but it may represent the mind with its sensations, feelings, thoughts, volitions, and other qualitatively defined states. This possibility will intrigue us later, for if it is accepted, the P-plane can be said to separate the external from the internal world. It is the locus of the contacts between the two, the carrier of sensations (more precisely, protocol experiences) that inform us about the external world.

3 *Varieties of Human Experience*

THE PRECEDING CHAPTERS have introduced the ideas of realms and domains of experience and have attempted to demonstrate the need for widening the meaning of reality. Now we look at the subjective nature of physical reality—a large, compound realm that includes primarily what we have called the see-touch or sensory domain, but also other domains, the contents of which have in common a clearly analyzable genesis and a unique epistemology. To do this we should understand the conventional analysis of human experience, so we may then focus on one part of it, the part that leads to physical reality. From the illumination of this subjectivity should emerge suggestions as to how alternate nonphysical realities may be interpreted and constructed. The details will be clarified by symbolic diagrams that clarify abstract relations.

The word "experience" is one of the broadest and most indefinite in our language. It shares this characteristic with its ancestor, the Latin verb *experiri*, which can be translated in many ways, among them "to try out," "to investigate," "to risk," "to try, in a legal sense," "to probe," "to learn," "to see," "to find," "to suffer," and even "to dream" or "to imagine." All these meanings are included in the current usage of the word "experience." To learn by experience implies an exposure to fact, often sensory fact, and the adjective "experiential" denotes something externally perceived or verified, in contrast to what is merely felt or thought or believed. But we also experience pain, suffering, a storm of ideas, a temptation, a desire, a doubt, the agony of making a decision. To collect all these strands of meaning we should use the word "experience" in a sense formulated by the philosopher William James and define it as "any item or ingredient within our stream of consciousness." This definition is loose but comprehensive, and we shall need one here that has maximum coverage. This large meaning of experience includes all sensory awareness as well as all phases of feeling, thinking, judging, willing. (The German language is somewhat unique in being able to capture all these aspects in one word, *Erlebnis,* the precise flavor of which is difficult to render in English.)

Included in the large meaning is the more limited connotation of "experiment." Its central meaning is "deliberate testing by an appeal to nature, an act in which man uses primarily his external senses." It excludes such components of awareness as mere thinking, feeling, and the creative act of theorizing. If one wishes to avoid ambiguity, one calls this latter, limited meaning of experience "empirical," suggesting an exposure to external fact, usually sensation, or the elaborate form of sensation called observation. This limited meaning is important for science, but it does not form the word's only or even its most important component. The word "empirical" comes from the Greek *en peira,* which means "on trial," and is therefore peculiarly well suited to designate the limited connotation of "experience." When referring to the latter, we shall use that word. For the present, however, we return to the larger significance of the term.

We have just presented a crude verbal classification of types of experience, a descriptive sort of classification. But there is another, perhaps a more profound distinction one can make: Experiences may be divided into a *cognitive* (the word comes from the Latin verb *cognoscere,* "to know") and a *noncognitive* class. The adjective "cognitive" has to do with knowledge; hence, cognitive experience is the kind that leads to knowledge—or better, perhaps—to understanding. The distinction at issue is vague at best, but as with many principles of classification, its vagueness does not render it useless. While science insists on precision and clarity of definition wherever these qualities are attainable, it also admits that there are "indefinables," "logical primitives," that defy exact definition. Among these are almost all experienced entities that enter the stream of our consciousness without our own contrivance. The things we see in the world, the contingent entities that confront us, are incapable of complete and exclusive definition. You might attempt to state all the properties you regard as composing the essence of a dog, including its having four legs, a tail, a bark. But this will exclude a dog that has lost a leg. On the other hand, our experience contains ideal entities like numbers, sets, towns, and nations, and these are terms that are capable of precise definition. The reason is clear: The latter are ideas, concepts, and images we ourselves have created; the dog is not. Let us note, then, that things we have constructed by ideal processes from cruder contingent kinds of experience are precisely definable, while the others are not. They can only be described, "hinted at," by some sort of pointing, by denotation, by what some philosophers have called "ostentative definition." We observe in this connection that the seen color green can be specified only by an ostentative definition, whereas its wavelength can be defined exactly. Or that the atomic weight of neon is 20.18 . . . with an unknown error. (This is because we ourselves have *chosen* the atomic weight of carbon as a standard and named it 12.00 . . .)

The distinction between cognitive and noncognitive experience is somewhat indefinite, like that between the seen green and blue of a rainbow. To clarify the distinction we must proceed ostentatively by pointing to examples. Obviously, feelings are not cognitive, nor is aesthetic enjoyment. Beauty, friendship, love, values, and to many people religious ideas like God belong to the noncognitive domain: as indicated, however, the distinction is not sharp. The artist sees a sunset and revels in its colorful beauty, whereas the physicist may be prompted by the same phenomenon to think of wavelengths, refraction of light in the atmosphere, and other matters that "explain" what is seen. To the artist the experience is noncognitive; to the scientist it becomes cognitive upon reflection. The word "cognitive" implies no value judgment—there is no way of deciding which is the more significant experience. That will depend upon the consequences: The artist may paint a picture of the sunset, the scientist may be led to a new conjecture about color, but these results are not comparable except in the subjective judgment of a given individual.

In many instances our language allows us to distinguish between the two types of experience. One can call one's dwelling a "house" or a "home." The former word designates the cognitive component of the thing in reference, the structure designed by an architect and constructed by a builder; the latter conveys overtones of feeling that add a noncognitive factor to the object. There are many such twins in any language—female parent and mother, retarded child and imbecile, personal attachment and love—that are often regarded as synonyms but that differ in the extent to which they carry noncognitive meaning.

The point of all this is simply that standard science deals with cognitive experience, leaving the other untouched. This does not mean, however, that the scientist as a person is immune to or disinterested in noncognitive happenings. Nor is the distinction absolute or permanent. For as science progresses, many noncognitive experiences will doubtless receive "explanations" and thus move, at least in part, into the cognitive domain. And it may well be that our claim that science deals exclusively with cognitive experiences is tautological, that we should look at the situation the other way around and say, Whatever science illuminates becomes cognitive.

Traditional philosophy makes a further distinction. It divides cognitive experience into "percepts" and "concepts." These words, too, need explanation and further elaboration, since their precise, literal meaning does not coincide with the intended meaning. A percept has reference to perception, to sensation, to the kind of awareness conveyed by the senses. A vision, a sound, a smell are clearly percepts. Concepts, or the other hand, are products of thought, imagination, and memory. The concept "man" is the abstract idea associated with a class of all men. Let us examine these two classes a little further, for they turn out to be of major importance for an understanding of science, and their treatment sets science apart from many other disciplines.

First we consider percepts. They are in the first instance deliverances of our senses. Their main property is a certain spontaneity, a "givenness"; their occurrence is somehow independent of our volition—we do not feel responsible for the fact that we see a tree. It is true that we may decide to look or not to look at the place where the tree can be seen, but the experience of seeing it when we do look is not evoked by us. This is often expressed by saying that the tree is "given in sensation," not merely thought. There is a philosophy called naive realism that takes the essence of nature, of the world, to be precisely what our senses reveal; It regards "being" as the sum total of all possible percepts or sensations. We need not concern ourselves at this point with the adequacy of this philosophy, although it will not escape the reader that the view involves difficulties because sensation depends on accidental circumstances (an object, for example, looks different under different illuminations; if our eyes were sensitive to X rays, our bodies would exhibit a different sort of reality than they now do). Furthermore, as we have seen, the perception of reality depends in large measure upon the culture within which one is raised. In spite of such vagaries normal science makes contact with the world ultimately through sensations. If physics, chemistry, or biology, for example, contained theories that did not somehow affect our expectations as to what can be sensed, or did not lead to predictions of specific sensations under nameable conditions, we would regard such theories as failures, as mere speculations.

But are sensations the only experiences nature presents to us when we are in a state of alertness? Are percepts the only ingredients of consciousness that come to us without solicitation? Before answering these questions, let us widen the meaning of "percept" and include in it sets of connected sensations, the kind of experience we call an observation, the result of an experiment—or, in more technical language, any form of "contingent information" about the world (the word "contingent," you may recall, means something extrinsic, usually unexpected, something that could not be established by thinking alone, something that requires "a look at the world"). Seeing an unexpected object is contingent; knowledge that $1 + 1 = 2$ is not.

In this larger context let us repeat the question: "Are percepts, sensations, the only experiences that assail us contingently?" The answer is clearly no, for there are introspective insights and intuitive apprehensions that share the spontaneity of percepts. A sudden unexpected recollection, the coming of a pain, the incidence of a mood—all these are "given" in a sense very similar to sensations. We therefore wish to include them within our class of percepts. To be sure, the physical sciences have ignored the results of introspection, and behavioristic psychology tries to explain them away. Yet there are many disciplines including some sciences (e.g., psychiatry, depth psychology) that take them as seriously as, or more seriously than, they do ordinary sensations. Their inclusion, however, falsifies the meaning of the term "percept"; for that reason we shall propose another word below.

Concepts, the label of the second class of cognitive experiences, denote the results of cogitation. The simplest view regarding their origin is the collective theory that a concept is related to a percept as a *set* is related to one of its members. The concept "man" is regarded as the set of all men. This pleases the logician because it makes matters very simple and allows the tools of modern set theory to be employed in an analysis of concepts. But there are some difficulties with this thesis. Science invokes concepts relating to entities that can never be a collection of perceptible entities or happenings. Nobody has ever seen, heard, or smelled an electron—electrons are much too small to display these sensory attributes—yet the concept "electron" is important in the current theory of atoms. In fact the electron, like many other scientific concepts, is somehow *postulated* vis-à-vis a certain collection of percepts.

The manner in which such postulation occurs is one of the important problems of modern philosophy of science. Let us here merely mention other scientific concepts whose genesis through postulation and not through setlike assemblage is clear upon a moment's reflection: in physics, all the invisible entities of the microcosm (atom and nucleus); in chemistry, the concept of valence, acidity, atomic number; in biology, organic form, evolution, heredity, life itself; in economics, concepts like market, gross national product, rate of inflation.

Some of these concepts may seem far-fetched indeed. However, the act of theorizing, assuming, hypothesizing—what we have called postulating—is very common, in elementary form, in almost every act of knowing. The experience of seeing a tree suggests at first blush a purely perceptory awareness, devoid of all hypothetical components. More careful analysis reveals the following: What sensation yields is a complex shape, an area of colors, gentle motion, perhaps a tactile impression of hardness, and so forth. But do all these bits of sensory apprehension, taken together, constitute the tree? Note that we endow the tree with an interior below its bark even though at the moment we see the tree, it is not in evidence; we assume it to be there. Note also that we assume that the tree has self-identity, permanence of existence: We take it for granted that it is there when we are not looking—in fact, when nobody is looking. Strictly speaking, therefore, we immediately endow our sensations with integrative properties that make them cohere, give them rational substance, and in doing so we go beyond the strict data provided by the senses. Arthur Eddington, in his book *The Nature of The Physical World*, goes so far as to describe the desk at which he writes: There are really two desks, the one that is purely given in sensation (the complex of rectangular shape, brownness, felt solidity, hardness, etc.) and what he calls the physicist's desk, the latter being an external object in three-dimensional space, having an invisible interior, consisting of invisible molecules and a permanent presence independent of any observer. This distinction between the two desks is perfectly correct. Let it be clear, however, that the common usage of the work "desk" never suggests the former, the mere complex of sensations, but always the latter, and that the passage from one to the other involves the postulation of certain ideal elements such as permanent presence. In a strict sense the desk is a creation of our minds. Later we shall say that it is "constructed" by us vis-à-vis certain perceptory experiences, and we hope the reader will recall the special meaning we give to the word "constructing," so as not to take it in the carpenter's sense. Construction describes a passage from the data of perception to the realm of concepts and ideas. This is extremely important in modern science, and its full understanding is indispensable. (In a wider sense construction can proceed in many other directions.)

The example of the desk is trivial, for the so-called ideal aspects are minimal, but it typifies a process that occurs in every recognition of an external object by our senses. The immediate data are augmented by an ascription of such nonsensory qualities as permanence, hidden structures, objectivity (i.e., we expect that other observers see the same thing, an assumption of which our senses could never inform us!). This special, trivial kind of "construction" is almost automatic in every simple act of knowing. It converts a complex of sensations into a thing. A good name for it is "reification," for the Latin *res* means "thing" and *facere* means "to make."

But the term "reification" must not be taken in its narrow, literal sense unless the meaning of the word "real" is expanded. Most people use it in its strictest sense, unaware of its wider implications, certain that it means something absolutely definite, indisputable, and ultimate. The term "reality" is derived from it; hence it, too, shares these fixed attributes. If you ask for the evidence of this state of affairs, the answer is likely to be "common sense."

But even a very unsophisticated bit of probing at once reveals difficulties in the common-sense view. Nobody will doubt that the desk before me is real, but is its brown color real, too? It is not a thing but an attribute, a property of a thing. Are all properties real? Do we reify the blue of the sky? We certainly do not make it a thing. In other instances there is a slippage of the word "real" toward "true," as when one says: the occurrence or the story is real. There are also entities that are not accessible to the senses—e.g., space and time, even "particles" too small to be seen. Most people would call these real although they are not directly reified from sense perception. At best they are deductions, derivatives from sense perception. All these and many other examples that will occur to you are instances in which the literal conception of the term "reification" will not quite do. This already indicates that even in the cognitive domain, in which physical science rules, caution is required in the use of the words "real" and "reification."

We encounter even greater difficulty in areas that are not strictly cognitive. We hear sounds, rhythmic notes, that may be regarded as real by only a slight loosening of that word's strict meaning. But the sounds compound themselves into music, which we enjoy. In a similar way we see the colors of a picture, a real picture in the limited sense of the word, but we also find the picture beautiful. The passage from perception of colors to the picture involves reification, but what about the further passage to beauty?

To use more extreme examples, we experience sensations in our dreams and reify them into objects and persons. In spite of this reification, we say that the objects and persons are not "really real." This silly phrase exposes the inadequacy of the word. Similar comments can be made about the experiences of hypnotized persons, of mediums, of matters felt in mystic ecstasy and in religious encounters or revelations.

In all these instances what is immediately experienced is an incoherent ensemble of sensations,* emerging and disappearing bits of immediacy, that cry out for meaning, order, coherence. What we called reification in connection with simple cognitive experience is an instance of a more general transition from a "rhapsody of perceptions"—to use Kant's term again—to something coherent, stable, and meaningful. The specific term "reification," characteristic of attainment of order in the beginning phase of most ordinary sciences, should therefore be replaced by one of the more general set: "organization," "synthesis," "stabilization," "transformation to stability," or "systematization." "Reification" is a specific instance of these. Henceforth, when speaking of the more general process of organizing perceptions, we shall use the word "organization." But at this point a problem appears on the horizon. Put in the form of a question it is this: If reification is the prevalent step toward the establishment of reality, do the others, covered by the word "organization," also lead to realities, perhaps of different forms?

Before turning to this puzzling enigma, we should analyze in detail the method of ordinary, cognitive science, for which reification is the gate to reality and truth.

Every act of organizing immediate experiences requires justifica-

---

* William James referred to this incoherent ensemble of sensations as "blooming, buzzing confusion." As part of consciousness, even sensations in dreams, hypnosis, etc., are not distinguishable from different sensations.

tion and explanation. In the case of reification, we test its legitimacy first in the simplest possible ways, to be followed by rational procedures that must be explained more fully. Let us now return to our major theme.

We verify permanence by looking repeatedly, objectivity by asking others to look, the presence of an interior by opening the thing, and so forth. In the more complex business of science, however, such tests are much more involved. How, for instance, do we verify the existence of neutrinos, the valance of atoms, the composition of a gene, the reality of black holes, the presence of a neurosis or mental illness? We will say more later about the ways in which the "construction" of scientific concepts is verified.

Protocols, Constructs, Observables, Systems

Throughout the preceding discussion we used the words "percepts" and "concepts," in rough accord with traditional philosophy. The inadequacy has become manifest, however; percepts are not what the word implies; a sensation is something modern psychology has shown to be a *highly involved response to external stimuli mingled with autogenic factors,* i.e., with activities initiated by the sensing organism itself. On the other hand, concepts are not abstract sets of individuals, as in traditional logic, but the results of human creativity, the sort of things called ideal constructions. It may be well, therefore, to introduce a terminology that, while somewhat novel, is free of the implications we wish to escape.

To repeat, the purpose of physical science is to organize, to make rational and meaningful, all cognitive human experience. Now the most incoherent part of our experience is formed by the contingent and spontaneous percepts that impinge on our consciousness. A mere record of them, no matter how complete, would not be science. Nevertheless, they are its raw material, the stuff that science attempts to make meaningful.

Ancient Greek scholars were in the habit of gluing on top of each completed manuscript, as its first page, a sheet of papyrus called a protocollon, a word that has come to us as protocol. Its literal meaning is "first" (*protos*) "glue" (*kolla*). This sheet carried an enumeration of all the items with which the book dealt, much like a table of contents today, but it lacked, of course, the rational connective tissue the book itself supplied. Thus the protocol of the ancient book bore much the same relation to the entire book that the percepts bear to the whole of conventional science. We propose to use the term "protocol experience" abbreviated to "*P*-experience," in place of the somewhat misleading percepts or sensations, words surrounded by a psychological nimbus we wish to obliterate. *P*-experiences need not be sensations or percepts in the usual sense. A sudden remembrance, the advent of a mood, the results of a questionnaire sent out by a polling agency, a census—all these are neither percepts nor sensations, but as starting points for scientific investigations they function as *P*-experiences even if the sciences of psychology and economics have not been as successful in explaining them as have been the so-called hard sciences in explaining simpler observations. We will let *P* stand for protocol, but if the reader prefers to limit his attention to the physical sciences, he may read it as "perceptory" or "primary."

As to the word "concept," we shall replace it by "construct" in order to emphasize the creative, ideal activity of which it is the result. It will be abbreviated to "*C*." Examples to be given later on will put flesh and bones onto this seemingly capricious terminology.

First, however, we introduce a schematic representation that may facilitate the understanding—or at least the discussion—of the relation between *P* and *C* experiences. We think of the plane of Figure 1 as the (two-dimensional projection of the) field of cognitive experience. Its edge is formed by *P*'s. We choose the word "edge," suggestive of a plane without depth because we do not wish to complicate our philosophy at this stage by intimating that there is something (like an unknown reality, a mind, a substance, or a god) beyond. Whatever we mean by reality, substance, and perhaps by mind and god must appear somewhere to the left of *P* among the valid constructs confirmed by the sciences and other disciplines.

To the left of the *P*-plane extends the field of constructs, of *C*-field, the meaning of which we shall now illuminate by simple examples. Insofar as one can speak of single constructs, they will be designated in the figure as circles in the *C*-field. The difficulty to which we are here alluding is one quite generally connected with ideas in contrast to things. Is the mathematical construct, the idea "number," single or multiple? Surely it can be split into individual numbers like 1, 2, 3, etc., and is in that sense multiple. Yet there is clearly a certain unity about the concept number, because there is a unique method of generating all natural numbers—namely, the process of counting. Treating ideas as individuals is like identifying clouds: It is sometimes possible, but very often not. In the examples that follow this difficulty is minimal.

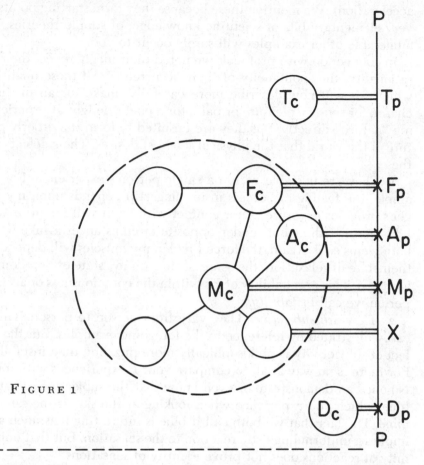

FIGURE 1

The word "temperature" means degree of hotness. Its elementary reference is to a sensation in our skin—for example, when I put my hand in a tub of water and find it hot. The experience alone is often regarded as unscientific; nevertheless it is the primary awareness from which a more refined meaning of temperature later arises. Let us designate it in the figure as a cross on the P-plane and label it T_P, the protocol experience of temperature.

Or consider force. Its P-aspect is the strain we feel in our muscles when we push or pull an object. In the figure we mark it as a cross and call it F_P. Other simple P's are the distance of an object from me, which I gauge when looking at it, the lapse of time I sense subjectively between two events, the color blue I see, the inertia I feel when I throw a stone, even the acceleration I experience in the push against the back of my seat when my car starts moving (resistance to acceleration). We mention these because they form starting points of very elementary bits of scientific knowledge, of simple theories. Innumerable other examples will surely occur to you.

In connection with protocols we noted their incoherence, their irrationality, the spontaneity of their occurrence. All these qualities, which we will now describe more carefully, make for an intrinsic charm, for sensory pleasure or pain, for a poet's delight at experiencing "nature" directly, but they are unsuited to form the orderly picture of the world that the physical scientist desires. Their defects are these:

1. They are unstable within a single person's experience. For example, the feeling of hotness in my fingertip depends upon my finger's previous exposure. After contact with ice it will record even a luke-warm bath as hot; under opposite circumstances a luke-warm bath seems cool. Again, the force I feel in my muscles will depend on their state of relaxation, the sense of time on my state of boredom or alertness. This unreliability of P's within the consciousness of a single person we call *instability*.

2. They are *subjective*. They vary from person to person. This is easily illustrated by reference to the foregoing examples. But the defect of subjectivity is philosophically more puzzling than instability. For there is no way at all to compare your P-experiences with mine. No one can demonstrate or even know that the subjective sensation blue, which you experience when looking at the sky, is the same as mine. The fact that we both call it blue is interesting inasmuch as it implies a uniform linguistic reaction to the sensation. But that uniformity of reactions does not prove identity of sensations.

This, incidentally, is one reason why some psychiatrists think that sensations in our P-sense are irrelevant, that the business of psychology starts with *responses* to P-experiences. This is a convenient supposition that is useful in certain applied sciences, even in so important a branch of modern research as information theory and computer technology. But because this view ignores the phenomenon of consciousness, ignores many important philosophic problems, and makes the passage from science to the arts more difficult, we will not adopt it here. Nor is there, as we shall see, any need to do so.

3. P-experiences are *qualitative* in the strict sense of that word: They do not carry numerical magnitudes; they are nonquantitative. I cannot attach a number to the seen color blue, or to the feeling of hotness in my fingertip, or to the felt interval between events.

It is precisely because of these three defects that scientists are forced to move away from the P-plane into the C-field. How do they make this passage?

The vehicle is measurement. To stabilize, to render objective and quantitative the P-experience of temperature, one employs an instrument called a thermometer, and one makes the reading of the thermometer *correspond* to the feeling in the fingertip. Logically, and indeed as an experience, the reading is different from the protocol of temperature. Nevertheless the word "temperature" is applied to both. This indiscriminacy is not without risk and has given rise to much philosophic confusion. We shall try to avoid it by calling the thermometer reading the *construct* temperature and denoting it T_c. As a moment's reflection will show, T_c is not afflicted by the three defects of T_p. It is stable, objective, and quantitative, for it is given by a number. To be sure, the number is somewhat arbitrary, for it requires the choice of a unit (degree). The need for such a choice is common to all measurements except one, the process of counting, which in our strict sense is also a measurement. In Figure 1 we symbolized the *measurement* of temperature T_p by a double line leading to T_c, the construct temperature. This measurement—a double line in our figure—forms a definition of the construct T_c.

Force, too, can be measured in several ways. The simplest is probably the one involving a dynamometer, which is essentially a spring and a scale on which the extension of one end of the spring can be determined. The other end of the spring is fixed. The agency exerting the force—e.g., a pulling arm—acts upon the loose end and moves it through a certain distance in the direction of the pull. The magnitude of the force is then the number of inches (or other length units) through which the end has moved. That number refers to the *construct* force, labeled F_c in Figure 1, and the measurement is again symbolized by a double line. Note again the nonidentity of F_p, the protocol, with F_c, yet both are called force.

At this point we need not spell out the details of measuring distance, time, the color blue, inertia, and acceleration. The instruments involved are, respectively, yardstick, clock, spectrograph, balance (since inertia is mass and is proportional to weight), and accelerometer. In every one of these instances except one the name of the construct is the same as that of the protocol. Only in the case of color do we note a difference: the C corresponding to color is a wavelength. To repeat: Every construct we have so far considered is the objective quantitative counterpart of a protocol experience.

Let us insert here a brief philosophic commentary, mainly re-

specting terminology. The great philosopher-physicist Percy Bridgman called the double lines in Figure 1, here dubbed M for measurement, *operational definitions* of the constructs to which they lead. He
believed, in fact, that every acceptable scientific definition had to be
of this instrumental type. His view is called "operationism." Most
philosophers feel that it overshot the mark, that even in science one
needs definitions of the noninstrumental, nonoperational type. But
more about this later. The instruments envisioned by the operation-
alists may not be of the simple variety used in the "hard" sciences;
they may be the act of questioning in psychiatry, or the tripping of a
lever that exposes food to a rat (operational definition of hunger in
psychology), the questionnaire sent out by social scientists, or the
counting of deaths in a society to "measure longevity." Indeed,
Bridgman included among his operations what he called paper-and-
pencil operations. At this stage, perhaps, the concept of operation
becomes so diffuse as to be meaningless.

At any rate there is a class of procedures, of which measurement is
an important example, which link the P's to the C's. We prefer to call
them *rules of correspondence* whenever the entire class is being referred to, which will not be very often. This term, introduced for this
particular purpose by one of the authors, is occasionally used in textbooks on philosophy of science. We mention it here because the act
of reification, which can hardly be called a measurement, is perhaps
the simplest rule of correspondence. In the fashion of the double
lines of Figure 1 it links a complex of sensations with an external object. In the figurative symbolism of our diagram it is a very short passage—e.g., from the complex of sensations, brown, rectangular surfaces, hard, etc. labeled D_p to the construct desk, D_c. It is thus seen
that there lie immediately in front of the P-plane innumerable constructs denoting things. Some philosophers would include them in P,
because the passage from the strict P to the thing it signifies is practically automatic (although it can be wrong, as in dreams and hallucinations). This would convert the P-plane into a thin sheet. Whatever
interpretation is chosen is immaterial for our purposes.

Ordinary sciences such as physics and chemistry get their start
after the occurrence of a P-experience and a measurement. Let F_p in
Figure 1 denote the sensation of a force, F_c the construct related to it
by a dynamometer measurement. In a similar way let M_p stand for
mass, M_c for the construct yielded by the use of a scale; A_p and A_c are
sensed and measured accelerations. F_c, A_c, and M_c are numbers in
terms of certain units. Now it happens, and this is perhaps one of the

miracles which makes physical science and the establishment of physical reality possible, that there is an invariable numerical relation between F_c, A_c, and M_c, of the form $F_c = M_c A_c$. A law of nature has been discovered! We have designated this rational, mathematical relation in the figure by single lines connecting the constructs, and we shall continue to use single lines for logical and mathematical relations between C's. A law, being quantitative, could never have been discovered among the qualitative P's. Here emerges the crucial importance of measurement. The law that has emerged here is known as Newton's second law of motion.

This situation, in which mathematical relations are discovered after the movement from P to C for a number of experimentally different P-items, prevails throughout physical science. A complex of related C's, such as those enclosed in the broken-line envelope, constitutes a *theory*. Some of the C's inside the envelope are left blank; they could be time, distance traveled, momentum, and others, in which case the theory would be called "Elementary Dynamics." We shall encounter many other examples for a variety of different sciences in subsequent chapters.

We have called the connection between P-facts and constructs "rules of correspondence." These are usually measurement operations, performed with instruments. A closer examination of them forces us to distinguish three types of rules. There is first the usual one, inherent in the examples we have discussed. It is called an operational definition of a construct—for example, the definition of temperature as a thermometer reading. However, there are two others, the first so universal and simple that it is rarely recognized or discussed. It is the one that takes us from individual sensation immediately, "instinctively," without reflection, to what we call the objects of the external world. It connects the individual sensations of the shape, color, size, hardness, smoothness of the desk before me—what Eddington would have called the positivist's desk—to the physicist's desk. It is a transition from an immediate welter of P-experiences to a consolidation in the form of objects, a consolidation that involves the guiding principles presented in chapter 5. The intervention of a rule of correspondence is required, for the two are not identical.

The conglomerate of sensations does not give the surface of the desk a rectangular shape; from where I sit it has a trapezoidal shape. This does not imply that the same sensations will occur when I look again, or that another person will agree with my implicit imputation of rectangularity of surface, of continued existence and permanence of the features I have constructed on the basis of my sensations.

The rule of correspondence that leads from the immediate complex of awareness, of *P*-facts, to an external object is *reification* (from the Latin *res*, meaning "thing"). It produces, as it were, the naive realist's world, which would be represented in our figure as a thin layer (thin because it is without rational relations in the *C*-field) adjacent to the *P*-plane. Within that layer objects are not connected by law; nothing can be said of them except that "they are simply there," constructed out of *P*'s. Reification differs from an operational definition because of its simplicity, universality, its uncritical character, its lack of an organizational structure and aim. It may be called the first step into science but is not alone able to build one.

The third type of rule of correspondence is more complex. It is a combination of an operational definition (double lines in Figure 1) and a further, logical relation (single line between two constructs in the figure). To see why it is necessary we note that a double line, which always originates on the *P*-plane, defines a quantity, a number corresponding to a sensation. In modern physics this is called an observable—more specifically, a quantitative observable. In Figure 1 mass, force, time, and distance are such observables. But every science contains more than observables; it bristles with entities like bodies of all sorts, solids, gases, liquids, and even onta, to which observables are *assigned*. In the see-touch realm these carriers of observables are themselves observable, but this ceases to be true in many realms. Atoms and so-called elementary particles are not directly observable, but some of their properties are. The same is true for the interior of the sun, the magnetic core of the earth, an electric field, a mental depression, and indeed the mind of man.

A similar complication arises when we define these entities, which are the carriers of observables. A material body is an object that has mass, size, shape, position in space, velocity, and many other observables capable of being operationally defined. An electron is that which has, or possesses, a specific mass, a charge, a spin—possibly a size. Thus, what we have denoted as an entity, an object, a system, something to which we assign existence in a more substantial sense than to an observable property, must ultimately be defined as "that which *possesses* or *carries* certain observables." The theoretical assignment of an observable to an entity or a system is a *logical* act, and logical relations are indicated as single lines in our diagram.

Thus, using a graph, an entity, or system—or, more generally, what we have previously called an *on*—would have to be represented by one or more ordinary rules of correspondence (double lines from *P*) defining observables that in turn lead by one or more simple lines to the entity in question. Thus, in Figure 2 operational definitions yield position (*x*), velocity (*v*), acceleration (*a*), and mass (*m*). The single lines assign them to a particle or a body. They say, in effect: a material body is *that which* has the observables *x, v, a,* and *m*. We shall call a definition of this complicated kind a complex rule of correspondence.

Figure 2 defines a special object, "*b*." However, the laws of nature, in this case Newtonian mechanics, are true for *all* objects that can be defined in terms of the same observables defining *b* in Figure 2. It is

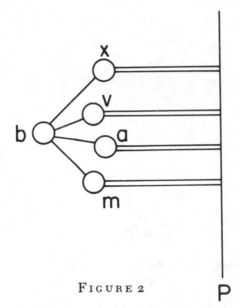

FIGURE 2　P

therefore not necesssary to draw one for each *b*. The laws concerning all *b*'s—in this case Newton's second law, force equals mass times acceleration—can therefore be represented by double lines from *P* to various observables, like F_c, M_c, and A_c in Figure 1, which are connected by the law.

We have given this very detailed—but always implicit—analysis of one aspect of the scientific method because it illuminates what we have called domains and realms, and furthermore because it may contain hints as to how experiences not yet responding to scientific treatment can be made scientific. Figure 1 represents a domain or realm—namely, one that connects the observables M_c, F_c, and A_c by a law of nature. The hint we receive is this: When examining phenomena as yet unilluminated by science within a certain domain of experience, look for suitable observables, then search for a law that connects them. The onta to which the observables belong (via a diagram like that in Figure 2) may be known to begin with, or they may emerge as suitable entities in the process of finding the observables and their laws.

4 The Meaning of "Truth"

PONTIUS PILATE, sitting in judgment on Jesus, asks casually, "What is truth?" (John 18:38). A more profound philosophical question has never been asked.

Science is often said to be the discovery of truth. Whoever is satisfied with this definition is not likely to become a successful scientist in fields where understanding, not merely the amassing of data, is required. For the word "truth," in spite of its air of finality, covers up many conceptual difficulties the thoughtful reader must face. There are, in fact, many kinds of truth, and they are in need of being distinguished.

The word "truth" is often used in isolation, without reference to anything else. There is the Platonic trio of virtues—truth, goodness, and beauty—all of them abstract ideals guiding the lives of men. But there is a difference among them: The last two, goodness and beauty, are properties or qualities of things and men; truth is not. It is a quality of statements, or propositions, or theorems, or laws, which are themselves statements. Many philosophers today ascribe truth only to sentences. Thus, a person can be *truthful* but not true. If one is said to be truthful, one is described as uttering only true sentences. This, then, is the first point to be made about truth: It is a property of statements.

But statements can be true in different ways or in different senses. A few examples: The simplest use of the word "true" occurs in the sentence: "A story is true." It affirms merely that the incidents related agree with observed facts, or, in our more specific language, that the description matches *P*-experience. It is in accord with what

[61]

one has perceived and does not go beyond this. Innumerable instances of this simple version will occur to you. Statements like "It is raining," and "The doorbell is ringing" are of this elementary type. Truth of this variety is called empirical truth; the word "empirical" comes from the Greek *en peira*, "under test," and the implication is a test against direct P-experiences.

However, the situation is not always quite so simple. Take the statement, "A molecule of water consists of two atoms of hydrogen and one atom of oxygen." You cannot verify this directly by looking at water. The reason is that oxygen and hydrogen atoms are constructs, related, to be sure, to P, but not part of it. Our statement is really a proposition concerning elements in C. But these are related to P by rules of correspondence, some of them operational definitions, and through these connections, the statement can be placed *en peira*. It is indirectly verifiable. (More about this in chapter 6.) Indirectly verifiable statements are also regarded as empirically true. Most scientific statements are indeed of this sort. To give a few examples: "In a vacuum, bodies fall with constant acceleration." "One body attracts another by a force called gravitation." "A blow on the body causes pain." "All living things die." Notice, however, that the sentence "All living things must die" would not be true in this empirical sense.

Most statements made in our daily life, if true, are empirically true, although there are instances where this characterization is merely implied or intended, but not actually present. In such cases difficulties arise that no amount of logical or scientific analysis can remove. That is to say, there are instances in which pretended empirical truth cannot be established. A witness reports what she believes to have happened. If she is the only witness, nobody can question her account. Yet she does not lie. On the other hand, her statements cannot be said to be empirically true. If I say, "I have a toothache," but the dentist finds that nothing is wrong with my tooth, I am making an empirical statement concerning a direct P-experience. But is this true? Or take the utterance, "I feel sad." There is no way in which this sentence can be verified by the methods of the previous chapter, the methods establishing physical reality. However, I know it to be true; hence my sadness belongs to a domain of reality different from the physical one. Similar circumstances surround many parapsychological claims. If I were to assert that every time I think of someone I love that person also thinks of me, the truth of the claim would be difficult to establish by the foregoing methods because of the turbulence of our lives and our forgetfulness. There are methods, in which such telepathic kinship could be statistically determined, but if they were found successful—i.e., if the number of coincidences exceeded normal expectation—there would still be many who would doubt the reality of such an effect. At any rate, it would hardly be regarded as physically real. It would most likely be regarded as a random occurrence.[1] These are examples showing that the decision as to empirical truth or falsity is not always easy or possible.

There are statements that fall into this category and still retain scientific interest and meaning. If I say, "Space is infinite," I am surely asserting something that cannot be directly verified in P. Nevertheless, the constructs space and infinity are clear, and they are somehow related to P, though in a rather complex manner. Figuratively speaking, the statement connects constructs far removed from P, yet distantly connected with it. Here the situation is such that there are, in principle, ways in which the empirical consequences of the statement can be checked, but astronomical observations at present neither confirm nor disconfirm it. And for that reason the statement is not empirically true at present, nor is it false. A similar analysis is to be made of the assertion: "Time has no beginning and no end."

To sum up the foregoing discussion: Empirical truth attaches to statements that are in agreement, directly or indirectly, with observation (P-facts). It depends on evidence external to the content of the statements themselves.

There are also propositions whose truth does not depend on empirical verification, which somehow carry it within themselves. Here are some examples. Simple statements, like $2 + 2 = 4$—indeed, all mathematical theorems—belong to this class. To be sure, one can verify that two plus two equals four by counting fingers; but somehow one feels that this is not necessary, that the result of adding two and two could not be anything but four. Propositions of this kind need not be quite so obvious. For example, the geometric theorem which says that the sum of the internal angles of a triangle equals 180°, while externally verifiable, would in this process cause greater difficulty than counting. There are others that are hardly verifiable by any empirical means at all. When I say $\pi = 3.14\ldots$, the measurement of the diameter and circumference of a circle would bear this out. But π has been calculated with the use of modern computers to 100,000 decimal places, and no actual measurements could possibly justify such precision. Another example of this type is the sentence:

"The function $1/x$ tends to infinity as x approaches 0." Note that infinity is an ideal concept, which cannot be reached except through vast extrapolation; hence this statement, too, is true in a sense different from empirical truth.

The preceding examples have this in common: Their validity can be decided by *analyzing* their contents; their truth resides within themselves. For this reason they are called analytic statements, propositions, judgments, or sentences, and their truth is *analytic truth*. The reader will observe that they carry within themselves a sort of necessity; one feels that they express something that could not be otherwise. But here a word of caution is in order.

One is not likely to doubt that $2 + 2 = 4$ until one begins to think more carefully about the objects to which this theorem applies. It is empirically true for fingers, apples, and all other objects whose identity does not change in the process of counting. But what about clouds in the sky, which merge and separate as the wind propels them? Do two ideas plus two ideas always make four ideas? Perhaps the sense of necessity we encountered is an illusion. The suspicion is reinforced when we examine our second proposition—the sum of the interior angles of a triangle equals 180°—which is true for triangles drawn on a plane sheet of paper but not for a triangle drawn on the surface of a sphere. Careful triangulation over large distances on the earth's surface cannot rely on its use.

Here we begin to see that analytic truth, once regarded as necessary truth, may not be true in an empirical sense at all. Scientists have learned during the last century that analytic truth is in every instance the logical consequence of certain axioms (i.e., very fundamental assumptions) and that it need not be applicable to the world at all. Thus, behind the statement $2 + 2 = 4$ there stand certain axioms of arithmetic, well known to mathematicians but beyond the scope of the present discussion, axioms whose prior acceptance entails all arithmetical propositions. These axioms themselves cannot be proved, but they are accepted, sometimes because of their simplicity, sometimes because they clearly lead to explanations of our experience, and sometimes for no stated reason at all. Our second example concerning the angles within a triangle follows from axioms first clearly stated by Euclid, as does the value of π to 100,000 decimal places. Finally, the limit of $1/x$ as X goes to infinity equals 0 depends on certain assumptions about the meaning of the term "limit" which has been clarified, again, only during the last century.

The story up to this point, then, is as follows: There are statements conveying empirical truth; their validity rests upon agreement with P-experience. A second class of statements expresses analytic truth; their validity arises from the fact that they are consequences of axioms, some of them very abstract and obscure, axioms whose symbolic place in our diagrams (Figure 1 in previous chapter) is far to the left in the C-field. Now the question arises: Are there propositions in which empirical and analytic truth are combined, propositions that follow from plausible axioms *and* that are true in the world? The answer is affirmative, and such statements are usually called scientific laws. (The word "law" is not used with precision in science—words like principle, theorem, formula often take its place.) They express a third kind of truth—namely, *scientific truth.* Here are some examples, discussed at length because of their relevance to our overall purpose here.

The great Copernican revolution occurred when the Polish astronomer Koppernigk published his famous *De Revolutionibus Orbium Coelestium* in 1543, the year in which he died at the age of seventy. Prior to this time the Ptolemian system of astronomy (Ptolemy, 2nd century A.D., an Egyptian astronomer who was born in Greece and worked in Alexandria) was accepted as descriptive of the motions of stars and planets. It assumed the earth to be the center of the universe and all planets and stars to move about it in a manner suited to heavenly bodies. Since the Greeks regarded the circle as the most perfect figure, the heavenly bodies had to move in circles. But a single circle for a planet would not do, since uniform motion upon that circle would cause stars and planets to revolve in uniform fashion in the sky about the human observer. Hence Ptolemy introduced two circles for each planet, one called "deferent," having its center upon the earth, and another the "epicycle," centered at all times on the periphery of the deferent, but moving with constant speed along that periphery. The planet itself revolved with its own constant speed on the periphery of the epicycle. By this composition of circular orbits the apparent motions of planets and stars could be accounted for. The number of epicycles required to reproduce the data known at the beginning of the 16th century was 83.

In the 16th century Copernicus noticed something very strange. While the planet Mars fairly described the visual path in the sky required by Ptolemy's assignment of circles, its large periodic changes in brightness were not compatible with the small radius of its epicycle. By careful thought and observation Copernicus discovered that a heliocentric system, in which the sun is at the center of the planetary universe, provides a simple description of the planets' apparent motion, for it gets along with only about thirty deferents, and furthermore it accounts in an approximate way for the changes in brightness of Mars. As a scientific discovery this was a noteworthy achievement, though it is eclipsed in brilliance by many earlier and later scientific accomplishments. Historians of science, however, assign unique importance to the Copernican revolution, not so much because of its scientific greatness but because of its philosophical and religious implications, which contradicted the doctrine of the Church. Persecution of adherents to the Copernican doctrine, of men like Galileo and Bruno, ensued.

We have presented this account, interesting as an episode in the history of science, as a preamble to discoveries that formed scientific laws in the modern sense. Copernicus' theory was inaccurate; it did not agree with the remarkably precise observations of the astronomer Tycho Brahe (1546–1601), who published careful descriptions of the motions of "777 planets and fixed stars." The originator of accurate laws was Johann Kepler (1571–1630), a young collaborator of Brahe, who formulated what we now know as Kepler's three laws of planetary motion. They are:

1. All planets move in elliptical orbits having the sun at one focus.
2. A line drawn from the sun to the planet, though changing its length, sweeps out equal areas in equal times.
3. The cube of a planet's mean distance from the sun is proportional to the square of its period of revolution.

To finish our story we must now pursue history beyond Kepler to Newton (1642–1727) and recall his momentous discovery of the simple and beautiful law of universal gravitation, which assumes that every particle attracts every other particle with a force proportional to the product of their masses and inversely proportional to the square of the distance between them.

Newton had no way of "deriving" his "law" in a strict logical sense. Observations were suggested to him, but his conception was a creative act, a leap from facts to a magnificent conjecture, inexplicable in simple scientific or psychological terms. As evidence we need only mention that in its strict interpretation Newton's law applies only to infinitesimally small particles, and these are ideal entities not encountered in the world of observation.

In strict logic, therefore, Newton's inverse square law is a *postulate* or an *axiom*. (We use the two words postulate and axiom interchangeably; they denote statements that can not be derived from other, more fundamental ones.) And now comes the major incident of our story: Newton showed by mathematical reasoning that Kepler's three laws are precise analytic consequences of his three laws. We thus see that Kepler's laws are analytically true.

But, as follows from the preceding, they are also empirically true, for they agree with Brahe's findings. Thus they are suspended between postulates and observations; they are doubly anchored in C and P. This illustrates the nature of scientific truth.

Our last example is chosen from biology. The beginning of the science of heredity was the discovery of Mendel's laws. They state, for example, that in cross breeding certain hybrid specimens of plants— for instance, the flower called four o'clock (*mirabilis jalapa*)—25 percent of the offspring will have white, 25 percent red, and 50 percent pink flowers. Mendel's laws, which he discovered in 1866 but which were not appreciated for several decades afterward, attributed to each plant certain characters or factors that he supposed to be the carriers of heredity. At this time they were abstract constructs which he postulated, and endowed with probabilities of distributing themselves in certain simple ways within the offspring. On the basis of these probabilities the physical appearance of the specimens in the filial second generation were statistically predicted, and the predictions were found to be true (for certain Mendelian characteristics). Thus, again, a set of scientific laws derived from basic assumptions (unproved existence of characters with determinable tendencies for combination) allows itself to be tested empirically in the appearance of concrete organisms.

This example is interesting in another way, for it shows how, in the progress of science, a set of postulates can be changed into laws derivable from more basic assumptions and then become the axioms of a theory. This is what happened in the science of heredity. Mendel's characters were shown to have actual existence within visible chromosomes; they are now called genes, and their distribution, i.e., the essence of Mendel's laws, can be derived in large measure from the chemistry and physics of the substance known as DNA. Here we witness again how truth arises in the double anchoring of a theory in C and P.

The axioms of the laws of heredity have thus become identical with those of the theory of chemical action. Two different sets of postulates have become one, and this always marks a triumph in the history of science. What happens here is depicted in Fig. 3. One might say that this amalgamation involves a further recession into the C-field, back to a more basic set of axioms. The reader will note that the complex of constructs farthest to the left, *a, b,* and *c*, forming the boundary of the C-field, are always axioms, for there is nothing to the left of them from which they can be derived.

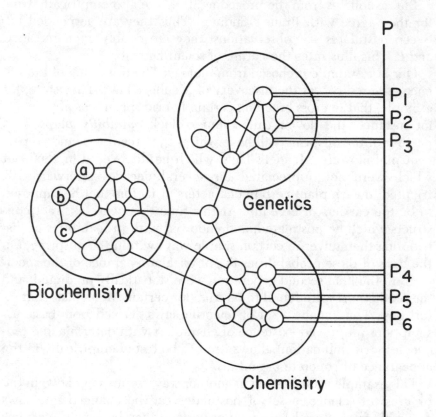

FIGURE 3

We have now seen the distinctions between three kinds of truth—empirical, analytic, and scientific. Now we present a challenge, which is to emphasize that a given sentence can be true in any one of these three senses and that the claim of a sentence may be ambiguous unless its sense of truth is specified. Many people accept the truth of the statement, "God exists." This may be analytic. If, for instance, we follow one of the "proofs" given for this proposition by the philosopher Anselm of Canterbury, we find him claiming this as analytic truth. For he says, "By definition, God is the most perfect being. If he were merely an idea in our minds he would not be the most perfect being, since he lacks existence. Therefore the very definition of God implies his existence."

The sentence may also be taken as claiming empirical truth, as it is by prophets and saints who have alleged to have seen and heard God's physical presence.

Finally, the proposition may be construed as true in the scientific sense. For it may be argued, first by way of basic premise, that the world requires a creator. This is clearly postulation, for it is also conceivable that the world has always existed. From this premise, however, one may derive a belief in God's concern for the world and, perhaps, for the affairs of man. If, then, this concern becomes manifest, perhaps in convincing responses to prayers, empirical verification is at hand, and the truth of the sentence becomes then scientific.

The reason why God is not generally regarded as a scientific construct is that both premise and empirical verification are open to doubt, the latter being highly subjective.

Henceforth, whenever we speak of truth, we shall mean *scientific* truth, the property of validated and accepted theories, which are logically or mathematically connected *constructs* having useful linkages with protocol experience via *rules of correspondence*.

We close this chapter by recalling a parable.

One of the oldest legends of our culture dates back to the era before the Libyan dynasties of Egypt, many centuries before the Christian era. It relates to the town of Sais, in the delta of the Nile, where a great temple had been dedicated to Osiris, the god of the underworld. Its ruins are still visible today.

It is said that this temple contained a mysterious picture that was covered by a veil and inscribed with the tantalizing words, "The Truth." Mortal man was forbidden to lift the veil, and the priests of Osiris enforced this statute with careful rigor.

A youth, dedicated to the discovery of truth (perhaps a scientist, if we may use a modern term), once entered the temple and saw the covered image. When he asked a guide whether he knew what was hidden by the veil, he received a horrified denial and an official account of the ancient law. The thoughtful youth left the temple that day, but an irresistible thirst for knowledge of truth forced him to return at night with intent at sacrilege. In the ghostly light of the moon he entered the hall of Osiris and lifted the veil from the image. What he saw nobody knows, but the legend insists that he was found near death, at the foot of the picture, by the attendants of the temple the next morning. Revived, he would not speak of his experience, except to regret it. His life thereafter was spiritless, his actions were undistinguished, and he sank into an early grave.

There the legend stands at the very beginning of our history, seemingly pregnant with significance, yet noncommittal, sphinxlike, foreboding human agony over truth, symbolizing one of the great and noble passions of man. It has not lacked interpretations: Some writers have made it imply the finiteness of the human mind, which cannot comprehend absolute truth. The German poet Schiller has given the story moral content, claiming that truth is fatal to a sinful conscience: "*Weh dem, der zu der Wahrheit geht durch Schuld, sie wird ihm nimmermehr erfreulich sein.*" Others have said that only God can reveal truth, and He will not be forced to it by human impetuosity.

We propose a different resolution of the ancient myth. It is that the youth of Sais, as he lifted the veil, might have seen engraved on the temple wall a message such as this:

Only a fool looks for truth in a finite formula; only a knave would want to acquire it without toil and heartache. Final truth is tantamount to stagnant knowledge; there is no substitute for self-correcting, progressing, ever-searching understanding. Dismiss your quest for truth in final formulation and embrace the greatest human virtue, called Unending Search for Truth.

Apparently the shock of this message destroyed a feeble soul that looked for truth by easy revelation.

Creed

Let us return, after portraying the meaning of truth, the scope of science, to the substance at the beginning of our discourse. We called science a *faith*, a restless faith, and we can do no better than present in closing some items, brisk and unconventional, that might serve, however imperfectly, as the creed of modern science. Here they are for you to ponder, to accept or reject.

1. We believe that the search for truth is a never-ending quest; yet we pledge ourselves to seek it.

2. We will not recognize or accept any kind of truth that pretends to be ultimate or absolute. We will consider and weigh all claims as provisional conclusions. If examination shows them to be stop signs on the road of inquiry, we will ignore them; if they are signposts, we will note them and move on.

3. We recognize no subjects and no facts that are alleged to be forever closed to inquiry or understanding; for science, every mystery is but a challenge.

4. We believe that new principles of understanding are constantly created through the efforts of man, and that a philosophy which sees the answers to all questions already implied in what is *now* called science is presumptuous and contrary to the spirit of science.

5. We are confident that scientific illumination can be made to penetrate not only the realms now affirmed as scientific, but also the shadowy regions that surround human consciousness, the essence of the mind, including features that are still obscure or occult and mysterious.

5 Guiding Principles in the Search for Scientific Theories

TRUTH IN SCIENCE, to recapitulate, arises in part from formal principles like those involved in pure mathematics, in part from its orientation toward P-facts. This latter source is called empirical verification of scientific theories; it is a process in which consequences analytically derived from the theory are checked against direct observational experience through experimentation. The former source, a set of formal principles, resides in a few very basic assumptions or conventions that have slowly evolved in the history of science. They are not imposed upon us by P-experience; though strictly independent of it, they serve to organize the experience. The words "assumptions" and "conventions," are perhaps too feeble to describe their power of organization, for they are more fundamental than most of the special assumptions made in any given science. Philosophers have referred to them as "rules of thought," as "categories," and as "metaphysical principles." Some philosophers, like Kant, have regarded a certain group of them as forever predetermined by the structure of our minds. We shall call them *guiding principles* and illustrate their role. After discussing them we shall deal with the equally important aspect of empirical verification, noted above.

There is a phrase, an aphorism, known to philosophers as "Occam's razor." Occam was an early philosopher of science who was among the first to recognize the importance of our guiding principles. His razor was not meant to shave in a concrete sense; its purpose was abstract. Occam's razor was a rule demanding that all inessential features be shorn off a scientific theory. In his Latin phraseology, *"Essentia non sunt multiplicanda praeter necessitatem"* ("one must not proliferate constructs beyond necessity," in loose translation). What that necessity is he does not make very clear, but his theory means simplicity, or adequacy, a minimal use of constructs, in the formulation of theories. Scholars of a later day like Mach (1838–1912), a physicist whose "number" is familiar as a measure of aircraft speed, alluded to the same feature of theories as occasioned by "economy of thought." Whatever their name, these guiding methodological principles can be spread out into a spectrum of separately nameable items, not independent of one another but useful for specific discussion. But first we should look at one or two examples to help us understand why guiding principles are necessary in addition to the P-experiences and the rules of correspondence introduced in chapters 2, 3, and 4.

Rules of correspondence, operational definitions, when applied to protocol data without further guidance, are arbitrary and do not lead to constructs which are unique. Take as a first example the operational definition of time, which involves an appeal to clocks. But there are many clocks, and time intervals measured by different devices are given by slightly different numbers. A pendulum clock, which depends on gravity, will assign different measures to the same interval if it is used in places where the force of gravity differs; a sun dial indicates a different time (solar time) than does the revolution of the stars (astronomical time). Today we regard as the most "reliable" measure of time what is called atomic time, a time based upon the vibrations of electrical charges within an atom. But even the scientist will find it difficult to define clearly the meaning of the word "reliable" in this context; it does not mean accurate or reproducible, for there is no reason why devices other than the atom cannot be made as accurate and reproducible as desired. Nevertheless, everybody acknowledges an improvement of "reliability" as one goes from pendulum time to solar time to astronomical time to atomic time. Brief reflection shows there is a guiding principle involved in this appraisal. To see it we recall here the most basic law of mechanics, perhaps the simplest law of nature. We know it as Newton's first law of motion: In the absence of forces, bodies move along straight lines, traversing equal distances in equal intervals of time. This law is not precisely true for solar time, but it is correct so far as we know for atomic time. The lesson we learn is that *the scientist chooses from amongst innumerable possible operational definitions, the ones that lead to the simplest laws of nature.* Nature does not force him to do this, but in his practice, he is committed, often unconsciously, to this maxim of simplicity. Hence, the double lines in our Figure are not entirely arbitrary, but are chosen because of certain requirements placed upon the constructs to which they lead.

Many other examples drawn from physical science illustrate this point. Consider, for example, the constructs, volume of gas (V), pressure (p), and temperature (T). There is very little leeway in our definition of V because that construct lies so close to the P-plane that it is given practically by reification. As to pressure (p), its operational definition may involve a manometer, or a barometer. When reasonable operational definitions of volume and pressure are accepted, T is still at the mercy of scientists, and here they are confronted with numerous possibilities. They can measure it by choosing among several devices, including a mercury thermometer and an alcohol thermometer, based on what is called a Carnot cycle, because a truly ideal gas does not exist.

Only by using the ideal gas thermometer does he obtain the relation $P \times V =$ constant $x\ T$ for pressures and volumes of a highly attenuated gas. For the others the product PV would be a very complicated function of T. Again, simplicity dictates the choice.

Another simple example touched upon earlier involves force, mass, and acceleration. Only if their operational definitions are chosen in certain ways does Newton's law, force equals mass times acceleration, emerge.

These examples were chosen from physics and chemistry, and one may wonder whether our analysis holds for the nonphysical sciences as well. Here it is to be noted that the latter deal with much more complex data than the former and therefore are expected to lead to more complicated laws and theories. But the tendency to define "observable quantities" in ways that will yield tractable constructs is clearly there. The plausibility in fields like economics, where such constructs as GNP, unemployment, economic growth, and inflation are important, the absence of known simple laws has made unique operational definitions of these terms impossible. Economists use, in fact, several different methods for computing such quantitative measures—that is to say, they use different operational definitions of them. Uniqueness can be established only when simple laws, or indeed any acceptable laws connecting these measurements, are discovered. Their absence is the true reason for the claim that economics, like psychology and other sciences, is at present an incomplete science.

A few such "laws" have been discovered. One of them arises in the science of psychophysics, which deals with the physics of perception. It states that the intensity of a sensation is proportional to the logarithm of the physical stimulus causing it. For instance, the sensed loudness of a sound or noise as measured in decibels is proportional to the logarithm of the number of watts per square centimeter of sound. Moreover, this is valid only for the middle range of sensations.

The guiding principle operative in all these instances has been called *simplicity*, a word that is admittedly vague and has been interpreted in different ways during the history of science. The Copernican revolution, which reduced the number of epicycles from the Ptolemeic 83 to 17, is an example where simplicity acted in its clearest, numerical form.

In admitting vagueness of the term "simplicity," are we violating the scientist's concern with precision? The insistence on precise definitions of *all* terms, though present among scientists working in certain specialties, notably in mathematics, is not expressive of a general goal of all science and is indeed impossible in any cognitive area, as modern logic has shown.

Let us give an example of the regulative power of the principle of simplicity. Long after Newton proposed his inverse square law of gravitation, an irregularity in the motion of the planet Mercury was discovered. Because of a certain precession—i.e., its entire orbit seemed to revolve about one of its foci—it did not exactly satisfy Newton's law. A mathematician proposed, and showed, that if the exponent 2 in Newton's law were changed ever so slightly—to something like 2.003, if our memory is correct—the precession could be accounted for. But this suggestion fell upon deaf ears: No astronomer took seriously the possibility that a basic law of nature should lack the simplicity, the elegance, the integer 2 conveyed.

Lack of precision is indeed a characteristic of all of the components into which we are about to divide the spectrum of guiding principles. Working scientists are not troubled by this; somehow they know instinctively what they mean. There has rarely been a controversy as to which of two competing scientific theories is the simpler, which of several basic laws is more general, which of several mathematical equations describing a given area of experience is more elegant and more beautiful.

Search for simplicity was a guide of all the originators of the epoch-making theories of modern physics. An example is a quote from a letter Einstein wrote to his friend Louis DeBroglie, one of the founders of the quantum theory (translated from the German; letter dated February 15, 1954, published in Annales de la Fondation, vol. 4, 1979)

I have long been convinced that one can not find the substructure [of matter] by constructing it from the known empirical behavior of physical things because the necessary jump in thought would be beyond human ability. I have reached this view not because of the failure of many years of effort (in your field) but also because of my work in the theory of gravitation. The equations of the theory of gravitation were *only* discoverable on the basis of a purely formal principle (general covariance), i.e., through confidence in the greatest imaginable logical simplicity of the laws of nature. Since it became clear that the theory of gravity was only a first step toward finding the simplest general field equations it seemed to me that this logical path must be thought through to its conclusion before one can hope to obtain a solution of the quantum problem. Thus I became a fanatical believer in the method of "logical simplicity."

Having dealt at some length with simplicity, let us introduce more briefly several other guiding principles that are equally effective in our discrimination between valid constructs (and their combinations, called theories). The second one is called *extensibility*. The function of this guiding principle can be illustrated by two historical examples.

The enclosure in Fig. 4 contains constructs connected by logical or mathematical relations (single lines) and is called a theory. Some of its components (three in the figure) have operational definitions that connect them with the P-plane (double lines). These instances in P, each of which may correspond to many qualitatively identical observations—e.g., the falling of any body near the earth's surface—are said to be *explained* by the theory, in this case Galileo's or Newton's law. Now, if there were another theory—in the C-field—that explained not only P_1, P_2, and P_3 but more instances such as P_4 and P_5, the latter would be regarded as more extensible than the former, and would be accepted as more *true*. In fact, the former would be rejected in part or as a whole.

Aristotle explained the fall of terrestrial objects by his famous theory of natural and violent motions. Natural motion, he assumed, portrayed the tendency of objects to seek their natural place. He recognized four earthly substances—earth, water, air, and fire—that were arranged in layers, with earth at the bottom and fire at the top. A stone fell in air and in water because it sought its natural place at the

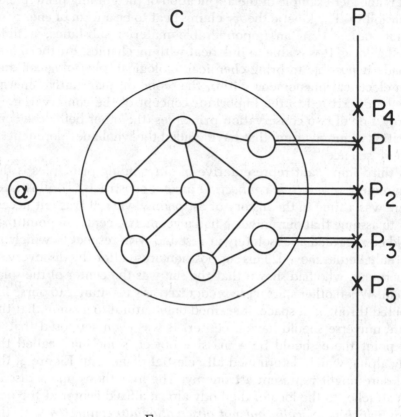

FIGURE 4

bottom. Similarly, air bubbles rose in water because they moved to their natural habitat, which was above water. Raindrops fell downward in air, but fire rose in air, and so forth. He thus presented a simple theory of the natural motion of ordinary objects. These motions, he assumed, took place without application of force. Forces, he thought, were able to change these natural motions, causing them to be "violent." The appeal of this simple theory is evident.

On the other hand, Aristotle's conjecture had one serious limitation, for it applied only to terrestrial bodies. Newton's law of gravitation, which superseded that of Aristotle, provided an explanation not only for the motion of objects on the earth but also for the motion of celestial bodies. It was more *extensible* than Aristotle's because it explained a larger range of observations—i.e., a larger domain of the *P*-plane.

The rejection of the caloric theory in favor of the kinetic theory of heat is another example of the application of the guiding principle of extensibility. The kinetic theory claims heat to be a form of energy of motion rather than an imponderable material substance, a fluid called caloric. It was able to link heat with mechanics. Furthermore, it made it possible to bring chemical, biological, physiological, and even electrical phenomena within the scope of quantitative energy calculations. It led to the important concept of the conservation of energy, one of two conservation principles (the other being the conservation of momentum) that have guided the whole development of modern science.

A third important representative of our guiding principles is the requirement of *multiple connection* among constructs. To illustrate: There was a time in the history of astronomy when it was felt necessary to assume that our universe had a geometric center, a point that could be regarded as absolutely at rest and with respect to which all heavenly bodies moved. This supposition arose after the discovery of Copernicus, who had shown that the sun was the center of the solar system. When other stars were recognized as suns similar to ours, distributed throughout space, it seemed only natural to assume that the entire universe should have a center. It was even supposed that at this point there should be a massive object, sometimes called the body alpha, which determined all celestial planets. In Figure 4, the enclosure might represent astronomy. The hypothesis under discussion attaches to the idea of the body alpha, a fixed center of the universe, *which is logically but not observationally connected* with the astronomical constructs. It remained isolated, singly connected outside the theory of astronomy, and it turned out later that it was never needed. It violated the principle of multiple connections and was therefore rejected.

Another more recent example may be cited. In Figure 4 the enclosure represents what physicists call the theory of beta decay, the emission of electrons by atomic nuclei. It did not contain the construct neutrino, that very small uncharged particle whose existence is accepted today. But earlier in this century it became apparent that, in order to save one of the basic laws of physics, conservation of energy, the emission of an electron should be accompanied by the emission of what the physicist Fermi called a neutrino. The construct was connected presumptively with the emission of an electron but attached to nothing else. It thus corresponded to the one extraneous circle α in Figure 4 and as such it was not accepted. But the compulsion to search for it led to its discovery. That is, the extraneous con-

struct was found to be logically connected by paths leading ultimately to observations on P with several other constructs in the
larger theory of nuclear physics. It became multiply connected. And
this fact caused the neutrino to be accepted as a part of scientific reality, as part of a valid theory.

Closely related to the principle of multiple connections, in fact
nearly identical with it, is one we shall call *logical fertility*. It requires that a theory shall have logical—and indeed, ultimately *empirical*—consequences. It must be possible to observe or do something different if the theory is true, impossible if it is not. If it entails
nothing, it is not accepted by the scientist or for that matter, by
common sense. Multiple connection ensures in part at least logical
fertility, for if a construct, or a set of them called a theory, is connected with other constructs, the connection is necessarily a logical
one, a relation satisfying our principle. It is introduced here as a separate guiding principle to rule out the following situation.

So far as we know, no material body in our world can move with a
velocity greater than that of light, c. The theory of relativity forbids
it. However, a slight extension of that theory wholly compatible with
the foregoing principles could be made to apply to entities moving
with speeds greater than c. If this theory is taken seriously—and it
stimulates many physicists to look for phenomena that might verify
it—it describes a whole world of onta (named tachyons) behaving in
accordance with consistent laws involving such concepts as mass,
speed, and energy, but entirely unobservable. The state of affairs
would correspond in our pictorial representation to Figure 5, where
a large, well-connected system of constructs, internally closed, fails
to stretch arms (double lines) to P. If this situation prevails, it contradicts fertility and will ultimately be rejected. Nevertheless, the very
possibility of formulating the theory, symbolized by the envelope in
Figure 5, induces physicists to look for the missing rules of correspondence. The stimulus is heightened by the formal elegance of the
theory, which increases the scope of special relativity.

Another example of a sterile, and therefore scientifically unacceptable theory, was the metaphysical philosophy of the great idealist philosopher Bishop Berkeley. His central belief was that all objects and happenings in the world, all human experiences, far from
suggesting external independent factors of reality, are thoughts in
the mind of God. The tree in front of your window is there because
God thinks of it. While this philosophy has a certain mystical appeal,
it violates the principle here under discussion, for it entails nothing

that could possibly be observed. It is symbolized by Figure 5. To be sure, we do not reject it as a philosophic or religious conjecture; it might indeed form the background of, and be compatible with, a scientific theory, but alone it is insufficient.

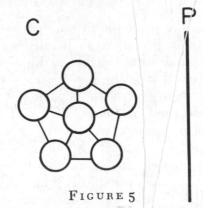

FIGURE 5

The next requirement, listed merely for the sake of completeness, is almost clear without being mentioned. We will call it *stability of interpretation*. It means that a given set of constructs, accepted as the explanatory counterpart of a certain *P*-domain, cannot be altered merely to suit different occasions. This principle has always been respected until, perhaps in very recent times, some writers on modern physics claimed—erroneously, of course—that an electron is sometimes a particle and sometimes a wave. We merely note here that such an assertion would violate the principle of stability of interpretation: Two different constructs are used interchangeably when convenience seems to require it. In fact, as we shall see later, an electron is neither a particle nor a wave.

Throughout the history of human thought about sensory reality perhaps the most important guiding principle has been *causality*. In simple terms, causality is the relation between cause and effect. We will discuss precisely what these are—whether things, events, or more elaborate constructs—more fully later. For the present, the ordinary notion of cause and effect as events taking place at specific times is sufficient. The principle states that a given cause always entrains a given, usually unique effect. Or in a more sweeping version, everything that happens has a cause. Implicit here is the understanding that the cause comes before the effect in time.

You may find it strange that we list causality as a guiding principle, a maxim we impose in the formulation of theories, when it seems that the causal relation is presented to us in our *P*-experiences. The view that it is thus presented, elaborated greatly by the British philosopher David Hume, is still accepted by many scientists and philosophers. But it seems that it is losing ground in the wake of the pioneering work of Immanuel Kant, who showed causality to be a regulative principle governing all of human rational experience. We shall here take it in this latter sense. Our assertion then takes this form: The constructs of science must be so chosen and combined as to exhibit in their description of temporal change the relation between cause and effect. The meaning of causality is, as we shall show in more detail, different in different realms of experience. In this respect it is more complex than our other guiding principles.

The last guiding principle is extremely vague but nevertheless highly important. It forms a bridge between science and art for it imposes the quality of *elegance* and beauty upon scientific theories. What is meant by elegance is difficult to define explicitly, but somehow every working scientist, and above all the creative genius who originates new ideas, new postulates, new fundamental equations, instinctively *knows* when a theory is beautiful. Men like Einstein and several outstanding mathematical physicists living today have gone on record to state what they mean by beauty of conception, by elegance of a formula (e.g., invariance). But it is not our intention to review their creeds—for creeds they are, like all the other guiding principles in this chapter. We do, however, wish to conclude with the emphatic affirmation that science in its present state of development is not devoid of aesthetic elements and that these appear to be of growing importance. Examples (invariance, symmetry) will be given later. No hypothesis or theory that does not satisfy the principle of elegance is taken seriously by science today.

To recapitulate: The guiding principles of science, the criteria employed in the acceptance as well as rejection of constructs and complexes of constructs called theories, have been loosely enumerated as *simplicity, extensibility, multiple connections, logical fertility, stability of interpretation, causality, and elegance.* They are not clear categories with rigidly defined boundaries but organically related requirements imposed upon the choice of constructs.

To this list of general principles we now add two specific ones. We have recognized space and time as constructs definable operationally and therefore directly related to the *P*-plane. It happens that they play a universal role in *all* branches of science, differing in this respect from all other observables, which appear in specific domains.

For this reason we introduce the following pair of further guiding principles: the use of time and the use of space in all scientific and presumably other realms.

It is true, of course, that such disciplines as statics, many branches of mathematics, and classifying sciences like botany do not make explicit use of time; this, however, implies that their results are true at all times. Hence our last principle is not violated.

The philosophically oriented reader will have noticed that our first set of guiding principles is similar to, but different in its constituents, from Kant's categories. But they are not offered as ultimate and unchangeable, which the categories were claimed to be. Kant called time and space "a priori conditions of the possibility of experience," an expression that comes close to our conception of them. But, unlike Kant, we do not regard them as immutable, not as having a single universal meaning. We shall indicate elsewhere that their precise meaning as observables may be different in different domains of experience.

Is our list definite and final? At one time this would have been affirmed, perhaps not for this but for some similar list of "categories" descriptive of the science of that age. They have been regarded as deeply embedded in the organization of the human mind, as external principles without which thinking was impossible. Scientists now take a more modest view and regard them as metaphysical rules that have slowly evolved in the successful pursuit of scientific inquiry during several millennia. Hence they cannot be guaranteed to be final. The future may add to them, or it may show some of them to be useless. If so, the changes are likely to be slow. For history shows that scientific theories live for decades, basic postulates like Newton's for centuries, and the guiding principles of science change within millennia. Their lifetimes seem to rival those of religions.

6 How Scientific Theories Are Verified

$\mathbf{A}$T THE BEGINNING of the preceding chapter we showed that scientific truth, that is to say the validity of an accepted theory, depends on two important kinds of factors: The guiding principles just discussed, and what we have called the process of empirical verification. While our immediate interest is primarily confined to "normal," or physical science, these two factors are crucial in the establishment of any theory relating to any kind of knowledge.

Take, for example, the game of chess. Its rules relate to possible moves of certain figures. These moves are constructs in the sense in which this word was used earlier, and there are correspondences relating them to P-experience, for the game can be played and observed. Furthermore, the rules are "consistent," and a careful analysis of this phrase will reveal that they satisfy most of the guiding principles of the previous chapter: One of these, causality, does not apply to them; they may be called elegant or pleasing; they are logically fertile insofar as they always lead to one of the goals—checkmate or stalemate. But we do not speak of their truth; they define merely a game. The evident reason for this is that except when we choose to play it, there is nothing in the world we know that behaves like the rules of chess; they are not verified in nature.

Consider a more specific and timely situation. We would like very much to know how our universe came into existence—whether it was created by a divine act, whether it evolved from some primordial state, or whether it has existed forever. Various conjectures elaborate each of these possibilities: biblical accounts of creation, the big bang hypothesis, and the so-called steady-state hypothesis. Each of these forms a coherent system of constructs that satisfy, at least in the belief of their adherents, all the guiding principles we have discussed. Some of their *observable consequences*, however, differ. The trouble is that the consequences are very difficult to observe and interpret, since they are indirect. Current theories concerning quarks, based on assumptions of simplicity and mathematical symmetry as guiding principles, lead to different types of quark behavior. But individual quarks have not yet been observed. The big bang theory of creation, despite its simplicity and cohesiveness, is difficult to check because we cannot go back in time. Astronomers at the moment tend to favor the big bang theory because one of its so-called "predictions," i.e., consequences, seems to have been observed: If the big bang took place, it should have emitted radiation, which in a finite and expanding universe should have diffused through space with a change in frequency that can be calculated. Such so-called background radiation has in fact been observed, and this is taken to be *empirical verification* of the theory.

We confront the caloric and the kinetic theories[1] of heat. The former was held to be true by physicists and chemists for a considerable period; its consequences among observations seemed to be confirmed. Then came certain discrepant discoveries, primarily those made by Mayer and Rumford.[2] These did not follow from the caloric theory. But while they contradicted the latter, they were in harmony with the consequences of the kinetic theory. Moreover it turned out that the observations previously "explained" by the caloric theory were also consequences of the kinetic theory of heat. Thus the former was rejected, but the latter was verified because it had a larger area of empirical confirmation.

Finally we mention that there are purely mathematical theories, such as abstract algebras, the concepts of which satisfy the guiding formal principles we have discussed (with the exception of those like causality and fertility that may not apply) but lack rules of correspondence with *P*-facts. These mathematical theories are by no means uninteresting; but insofar as they proclaim only formal (analytic) truth, they are called purely formal theories. These are confined to mathematics and to logic. They may even be useful in a prophetic sense; for very often in the history of science formal theories have acquired rules of correspondence and become applicable to the world, as the kinetic theory did when Rumford bored his cannons. Our interest is in scientific theories in the normal sense—that is, in nonformal theories requiring empirical verification.

Before we proceed to our analysis of how the process of empirical verification works, two general comments are in order. The first concerns terminology. In accordance with general—at any rate with desirable—usage, we shall speak of an unverified system of constructs as an hypothesis. Thus empirical verification converts an hypothesis into a theory. Secondly, we wish to forestall the impression that empirical verification, which is treated here as the second factor constituting a valid theory, is of secondary importance; it should be regarded as coequal with the guiding principles. There is in fact a widespread though waning philosophical view, occasionally held by scientists, which regards conformity with P-experience as the only important criterion of scientific theories and degrades the formal principles to mere matters of convenience or accident. The name of this philosophy is positivism, also labeled strict empiricism or inductivism. We claim that the view given in this book is a more balanced one and is in the ascendency.

In terms of our previous figures, the process of empirical verification is represented by a circuit that starts at a point P_1 in the P-plane, extends into the C-field, passes through a segment of it, and returns to the P-plane at P_2 (see Figure 6). Examples should help you envision this symbolism. For instance, the theory of planetary motion is summed up in Kepler's laws. They were, to be sure, "suggested" to him by Brahe's careful observations. Unexplained observations are often the starting points for the formation of conjectures: Genius

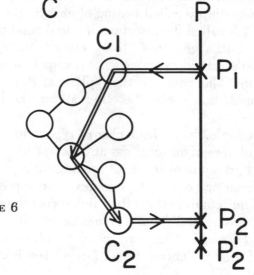

FIGURE 6

somehow performs the leap from P to C in ways that are incomprehensible to the ordinary scientific mind. But without further tests, i.e., without empirical verification, the conjecture remains an hypothesis lacking the rank of a theory. What then did Kepler do in order to promote his hypothesis to a theory? Let us consider in particular his second law, which says that the radius vector of a planet as it moves in its elliptic orbit sweeps out equal areas of the ellipse in equal time. Verification requires three steps. First, Kepler sifted from Brahe's data the distance from the sun, r, of a planet, say Mars, and the velocity v in its orbit. These data form the point P_1; ascertaining the corresponding values of r and v involves the rules of correspondence, which are represented by the upper arrow (double lines) to the left in Figure 6. Kepler had now arrived at C_1. At this point he employed the second law, which says in effect that r times v is a constant. Therefore, he reasoned, since at some other time, earlier or later, r was different, v must be different also. Thus for a new r the law "predicts" a new v. The word "predicts" is used in this manner by scientists. It does not necessarily mean that the "predicted" situation lies in the future; it might be in the past. This reasoning is symbolized by the lines from C_1 to C_2. It represents a movement among constructs and lies wholly in the C-field. At the point C_2 Kepler used once more the operational rules that link r and v to astronomical observations. He determined from Brahe's data whether v at the new r had the predicted value. Since it did, his hypothesis was confirmed.

There are common names for the three steps of this process of confirmation, often simply called testing of an hypothesis. The passage from P_1 to C_1 is called the insertion of initial conditions into the theory to be tested. Passage from C_1 to C_2 is simply the application of the theory. In the so-called exact sciences it usually involves the solution of equations. The return to the P-plane at P_2 is called a prediction. If the prediction meets the observations, the hypothesis is verified.

For another example, drawn from chemistry, consider Boyle's law: pressure times volume (of an ideal gas at constant temperature) is a constant.[3] Here P_1 designates the joint readings of a pressure gauge and a device measuring volume. C_1 denotes the computed values of pressure and volume computed via the rules of correspondence. C_2 is another pair of pressure and volume values having the same product as before, and P_2 means the corresponding observations of the meters. Again, if they are encountered, Boyle's law is confirmed (or tested or verified).

The process of empirical verification is always a circuit through the C-field starting and terminating at two different points in the P-plane (we shall call this an open circuit). It is important to realize this, for a theory is never confirmed by a reference to a single P.

The same kind of circuit is sometimes employed to verify rules of correspondence when a theory has been tested and is regarded as true. Suppose, for instance, that a man observes a luminous object in the night sky. Being an amateur he does not know whether it is a fixed star or a planet or, perchance, even a rocketship under power. He is in doubt with respect to the passage from P_1 to C_1—P_1 corresponds to position and velocity of the object, which we assume he has observed. He now has P_1 and first assumes that the object is a planet. This provides the first tentative step to C_1. He then uses the laws of planetary motion and arrives at C_2 which denotes position and velocity of the mystery light at a later time—say, the next night. He looks for its appearance at that time, and if he finds it, he concludes that his *interpretation*, i.e., the rule of correspondence he employed, was indeed correct. If not, he uses the other interpretations and tests them in a similar way. The open circuit of empirical verification thus serves as a test for any one of the three steps it involves.

On a more philosophic note, remember our earlier example of the desk. The claim was made that it—that an external object—is in the strictest sense a construct whose validity, in this case called physical reality, needs to be verified. This is done by use of a circuit of the type just described and in the following manner.

I am confronted with a set of immediate sensations (P-experiences), including the visual impression of brownness, rectangularity, shape, and size. I also have the tactile sensation of hardness and rigidity and perhaps a muscular awareness of its weight as I attempt to lift the desk. But all of these taken together do not strictly constitute the external object, which is assumed to have an interior that is not seen, is thought to be composed of atoms so small that they can never be seen in ordinary light, and is endowed with permanent self-identity—namely, the property of being there when I am not looking, indeed when nobody perceives it. All these latter properties, which transcend the P-plane, need to be tested. These are the tests:

I first proceed, in the face of the mentioned P-facts that confront me, to postulate the entity desk, endowing it with all the unobserved properties just listed. This postulation involves a rule of correspondence we have called reification: It amounts to the passage from P_1 to C_1, the latter being the construct "desk." In view of what was said previously, this C_1 should lie very close to the P-plane. Then I reason: If the property of permanent self-identity is valid, the object must appear when, after turning away, I look again. Equally important in testing for self-identity is the conclusion of "intersubjectivity," often called "objectivity," which demands that the object should be visible to other persons as well as myself. If it has an interior, it must reveal itself in terms of certain P-experiences when I open it to view; if it has atoms, then certain operations known to physicists and chemists must have predictable, observable effects, and so on. All these if-then propositions are logical passages from C_1 to C_2. The final step from C_2 to P_2 involves ascertaining the predicted effects. The existence of the desk, its so-called reality, is verified if they are at hand.

Of course we do not perform these tests every time we meet an external object, and we may even tend to regard this sort of discussion as trivial and needless. Nor do we follow the three steps here cited consciously. We perform them automatically without ordinarily being aware of them. The fact is, however, that we learned the process in our early childhood. A very small baby losing a toy will not search for it. It has not learned to reify. At a certain age, however, it stretches out its hands for things it has lost, making a primitive attempt to verify a feeling that the object must be there. For the baby the age of reification with attendant tests of its validity has begun.

But there are circumstances in which even mature persons more or less consciously fall back on the processes described; this happens when there is doubt whether a set of impressions, clear and vivid like the sensation of real things, refers to actual objects or is made up of hallucinations. Here the test of objectivity becomes explicit and important. If it fails, the physical scientist dismisses it, but the psychiatrist may see significance in the experience and try to understand it by different rules. The latter may then even assign to the phenomenon a nonphysical reality.

Before concluding our discussion of the process called empirical verification we must confront two additional important questions. First, does a single instance of verification suffice to confirm a conjecture, to convert an hypothesis into an acceptable theory? The general answer is no. What happens is that one gains more and more confidence in a theory as the instances of its verification increase in number. The story is that Newton conceived the law of gravitation when, resting under an apple tree in his uncle's garden in Woolsthorpe, he saw an apple fall in what seemed to be uniformly accelerated motion. Had he waited for another apple to fall, the growth of confidence in his conjecture would hardly have increased, for one thing because the second circuit of empirical verification was too much like the first. But when he applied his ideas to the motion of the moon about the earth and—after many vicissitudes—found that his predictions verified his theory's validity, his confidence in its truth increased significantly, and after several other tests it became generally accepted.

Science has never seen fit to specify the precise number of verifications a theory must endure to be accepted or, as the language has it, to be regarded as true. Nor, it must be reiterated, is its truth permanent and unchangeable, since the possibility of disconfirmation or falsification by later tests is always at hand. Somehow scientists develop a measure of assurance and unanimity regarding the validity of theories which defies precise analysis—to define validity is in some respects similar to defining beauty.

Philosophers have occasionally been dissatisfied with this looseness in the meaning of truth. They prefer to speak of the probability, not the truth, of a theory and to measure this probability by a number of successful confirmations. Scientists generally reject this alternative. They point out that if the probability is defined as the number of actual confirmations divided by the number of possible (but not performed) confirmations, that quotient is always zero, since the numerator is necessarily finite and the denominator is infinite. Furthermore, a single failure of the process of confirmation makes the theory false in the eyes of the scientist, while the probability would change inappreciably no matter how it is computed.

Finally there are instances in which a single act of confirmation can in fact establish a theory. These instances are called crucial experiments and involve observational results so striking that nearly all doubt regarding an hypothesis is stilled. An example is the prediction that a positron (an elementary particle like an electron but carrying a positive charge) should exist, made by Paul Dirac in 1928 on the basis of a mathematically attractive ("beautiful" or "elegant" perhaps are better words) equation. The positron was discovered by Carl D. Anderson in 1930, and Dirac's theory was at once acclaimed as true. Anderson's observation was a crucial one, unique, creating a measure of confidence not expressible by any assignment of probabilities.

The second fundamental question concerning the process of empirical confirmation concerns the closeness of the agreement between prediction and observation necessary for the circuit to be regarded as successful. Suppose a theory says that a radar beam sent to the moon and reflected at its surface will return after 2.58 seconds. If it is found to return after 2.56 seconds, does the slight discrepancy count as falsification of the theory of radar propagation? Clearly, many circumstances that complicate the observation—inaccuracy of instruments, errors on the part of the observer, uncontrollable vibrations of apparatus—must be taken realistically into account. In terms of Figure 6, P_2 as predicted will never precisely coincide with P_2^1, the protocol event actually experienced. Our question can thus be paraphrased to read: What is the maximum tolerable distance between P_2 and P_2^1 which still counts as successful verification? The answer is easy to state: The distance is the "probable error."[4] This convention is not entirely arbitrary; mathematics gives good but not incontrovertible reasons for its acceptance. The choice of probable errors has many of the philosophical properties of our guiding principles. We have mentioned it here because it emerges in the process of confirming theories and operates at the very boundary of P and C.

The foregoing considerations should be of special concern to workers in parapsychology. Their results, their P-facts, are first accompanied by a sense of surprise, even amazement, which can be so overpowering that other elements of the process of verification are ignored. One experience provokes an exclamation of *eureka*. This is the attitude of most amateurs. Trouble arises from the fact that a single instance, or even a few instances of the same kind, do not allow a discrimination between their being chance occurrences and the manifestations of a scientific law. To remove this uncertainty they arrange matters so that the same or similar results can occur. In some important instances (e.g., psychokinesis with random generators), this happens, with slight differences in their results. Assuming that the distribution of these outcomes follows a normal statistical law the researchers then apply the so-called binomial formula[5] and conclude that the outcome of all repetitions cannot be due to chance. This is, in principle, an important part of the process of verification.

But in essence it is a frequent repetition of Newton's observation, as if, lying under the tree, he watched an apple fall many times with the same acceleration. It stabilized the occurrence of P_1 in Figure 6, but never attempted to arrive at P_2. To do this one has to introduce a theory, establish a lawful domain. One has to quantify P_1, say ESP, in forms of observable constructs. This is done by researchers in this field when they regard the experience as thought transfer and assign to it the quantitative measure called relative frequency of occurrence. But here they usually stop: they rarely look for other observables in the realm of psychology with which frequency of ESP occurrence is lawfully connected. This would make possible a passage from P_1 to P_2 in the figure and presumably others, thereby initiating a scientific domain of parapsychology. One might suspect that ESP, clairvoyance, or other mystic experiences such as those related (but uncoordinated) by Eileen Garrett could ultimately be coordinated into a scientific domain.[6]

7 *Reductionism in Physical Science, I*

THE QUESTION MAY BE RAISED as to why we place a chapter on reductionism at the center of this book as though this topic were crucial to its purpose. But this, in fact, is the case.

The concept of reductionism is one that has led to tremendous traps and errors in science and philosophy, and it should be abandoned if further progress is to be made. To grasp this, we need first to understand the concept's historical and traditional use and the language that has developed around it; we can then see where it must lead and develop other concepts and language essential to the future of science.

We have now dealt with a variety of experiences and affirmed the need of describing types of reality that differ from the physical norm; thus arises the important question: In what sense are these different? Are they related to one another in the same way in which different physical theories, i.e., physical interpretations of reality, are logically connected? The latter are often said to be "mutually reducible." The meaning of that phrase is vague, and we must clarify the meaning of terms like "reducibility" and "reduction in the physical sciences" before proceeding beyond their boundaries to inquire, eventually, about the cohesion of different nonphysical realities. We find, unfortunately, a staggering variety of meanings, misconceptions, and indeed contradictions connected with the term "reduction." We take the physical sciences here to include biology, physiology, and chemistry, and the apex of a certain version of reductionism is reached in certain solutions of the the mind-body problem, which involves all of them.

The primary etymological meaning of the verb "to reduce" is clear: The Latin verb *reducere* means "to pull back." Thus, if one thing or one idea reduces to another, it is implied by it. But there was also a secondary meaning in Latin—"to save." If one idea seemed strange or unacceptable, its meaning could be saved by reducing it to a more familiar or a more acceptable one. The Romans associated various secondary meanings with the word: *uxorem reducere* meant to bring a rejected wife back home; *aliquem de exilio reducere* meant to allow a man to return from exile. All of these meanings have faint reflections in the modern scientific and philosophic use of the word.

The history of scientific reductionism dates back to the earliest writings in the field. In the beginning the entities subject to reduction were simply *things* (the Greek atomic theory), then *ideas* (heat is reduced to motion of particles), and finally elaborate theories involving ideas and things (the mind-body problem is sometimes solved by reducing the mind to parts of the brain). The atomic theory, which analyzes complex bodies into elementary ones, is the oldest and simplest of all reductivist attempts. Its origin is found both in the Orient and the West. The Hindu philosopher Kanada (500 B.C.), who preceded the first Greek atomists Leucippus and Democritus, practiced reductionism by claiming that all material substances are aggregates of primary particles, the smallest of which are invisible. "The mote which is seen in the sunbeam is the smallest perceptible quantity." It must be composed, Kanada reasoned, of what is less than itself, and this likewise has smaller parts.

This again must be composed of what is smaller, and that smaller thing is an atom. It is simple and uncomposed, else the series would be endless, and were it pursued indefinitely there would be no difference in magnitude between a mustard seed and a mountain, a gnat and an elephant, each alike containing an infinite number of particles. The ultimate atom is therefore simple.[1]

The reasoning is unique and beguiling; it concludes with the assertion that the "ultimate atom" is the sixth part of a mote visible in a sunbeam.

Some fifty years later Leucippus and Democritus gave a similar though less graphic account of atoms and molecules, to which all materials should reduce. Less sophisticated was the earlier reductive view of Thales (600 B.C.), who thought that water was the substance that forms the constitutive core of all complex matter. This kind of material reduction was practiced until the beginning of the present century, when radioactivity was discovered, when particles had their identity changed, and when simple material reductivism had to be abandoned. Nevertheless, these superannuated habits of thought remain.

But even in the science of ancient Greece the ultimate residuum of the process of reduction was not always matter. One of the simplest prototypes of the reducibility of *ideas* was offered by Parmenides (500 B.C.). His reliance on observation was minimal, but his reductive reasoning was keen. Convinced that something must exist because it was the subject of thought, but refusing to admit that "it" was as complex as matter appeared to be, he reduced its essence to what he called *einei* (being, existence), endowing it merely with ideal qualities. The senses he took to be false witnesses, *kakoi martyres*, which made being more complicated than it is. As an avid reductionist he deprived his idea of being of all perceptible properties and left it only with the quality of filling space. The *einei* thus became the *pleon* ("the full"); he claimed that being and filling space are synonymous. His final reductive act was the most radical: Since being and filling space are synonymous, there can be no vacant space. If empty space exists, it has being and therefore fills space. This penultimate step precludes the possibility of motion, for motion is the passage of part of being from where it is to where it is not, and since empty space does not exist, the condition "where it is not" is illusory. Thus coexistence and succession are nonentities, fallacies of the senses. The "existing" is eternal, uncreated, indestructible, unchangeable, and homogeneous. Parmenides pictured it as a sphere.

Parmenides made the first recorded attempt to subjugate the multiformity and flux of sensual experience to one supreme logical thought whose power flows solely from freedom of internal contradiction.[2] It was the first and perhaps the most extreme form of ideal reductivism. It should be clear that both examples, of elementary material and of unrestricted ideal reductivism, ignore the epistemology sketched in earlier chapters, disregarding the guiding principles as well as the need for empirical verification. Reductionism in modern science can be closely aligned with that epistemology. (In our treatment of reductionism, we are greatly indebted to the penetrating writings of Arthur Koestler, even though our examples are mainly drawn from physics. His books, *The Ghost in the Machine* [Hutchinson, London, 1967], *The Case of a Midwife Toad* [Hutchinson, London, 1971], *The Roots of Coincidence* [Hutchinson, London, 1972], and particularly *Beyond Reductionism* [Hutchinson, London,

1969] have paved the way toward conclusions very similar to those we are about to reach.)

The word "science," in contrast to its counterparts in other languages such as the German *Wissenschaft*, which has a wider scope, designates a very specific way of organizing human experience. Since the most basic empirical science is generally supposed to be physics—a subject that must be taken in its large sense to include chemistry, astronomy, physiology, and important aspects of biology—the coherent structure of theory or explanation to which it has led at any given time is called physical reality. The meaning of physical explanation, the processes leading to valid theories—i.e., the establishment of physical reality—is uniformly accepted by working scientists even though the details are, and should be, continually subjected to philosophical scrutiny. A physicist may be a materialist or an idealist, an empiricist or a rationalist, a monist or a pluralist, a person of religious faith or an atheist: The method of research among all these is astonishingly uniform. The essential features of the scientific method leading to the formulation of physical reality are those we presented earlier.[3]

In its most general sense, which we shall see is not very precise, modern reductionism is a composite of two features characterizing and providing the basis of the scientific endeavor. Neither will be accepted by strict empiricists, who regard science as a catalog of facts and their endeavors as attempts to complete the catalog. To them, their subject is very much like a picture puzzle that can be completed when all pieces are put in their proper places. But science is not a two-dimensional game. It has, as it were, a third dimension, which forces the investigator to probe below the facts in the fertile ground of theory, the region of what we have called constructs, whose concatenation binds the surface facts together and allows them to grow.

We can find an analog for the development of science in a growing crystal. A liquid substance is placed in a vessel at a temperature above its melting point. Its molecules are in disorder; science has not yet arranged them in patterns. But as the temperature falls below the melting point, crystallization sets in. The molecules, held together by unseen forces, arrange themselves in a regular order, crystal structure appears, and often a beautiful configuration emerges and grows. This growth is limited only by the surfaces of their vessel; in an infinite volume growth never ceases. Ever new portions of the liquid are transformed into a solid structure that ultimately fills all space. Science grows like a three-dimensional crystal, and it will never be completed. This analogy also illustrates the two features of the doctrine of reductionism. The first is the temporary character of science, its incompleteness at any given time. The volume of the crystal is finite at every instant of its infinite growth. There are times and temperatures at which parts of it will grow in abnormal patterns, and then, usually quite suddenly, the abnormal part will accommodate itself to comply with the norm and then grow in the regular way.

There is another way of describing the temporary character of scientific truth, its continual need for refinement, extension, and at times rejection followed by new research. We have called scientific truth "asymptotic truth,"[4] the light at the end of an infinite road of discovery, an ideal probably not within human grasp. Part of the scientists' creed is that scientific endeavor, as it alters and enlarges reality, introducing it perhaps into nonphysical domains, is not a random search but an approach to an ideal. The word "creed" is used here deliberately, for science, too, has its articles of faith.

A strange terminology has spread among some philosophers. They speak of the history of science as a series of revolutions, as though the abandonment of one theory in favor of another were a unique, uncommon, and upsetting occurrence. The inappropriateness of this usage should be evident, for the "revolutions" are continual, they are zigzag parts of an asymptotic movement that never subsides. At any given time there is a prevalent scientific view (sometimes called a paradigm, which means model, exemplar, or prototype and is normally used to designate a typical grammatical sequence of forms such as *amo, armas, amat*), but unless the prevalent view is dogmatized, it is viscous but fluid (i.e., not given up without resistance).

The second feature of interest in connection with the reducibility problem—in fact, a feature located in the center of interest—is the metaphysical principle earlier called "extensibility of constructs." As we have indicated earlier (see Chapter 5) it is usually employed as a criterion for eliminating useless theories; it demands that when two theories compete in explaining a certain set of observations (P-facts in our early diagrams), the one with the wider range is to be retained. This may happen in three ways: (a) one of the theories will be rejected; (b) with a reinterpretation of constructs in one or both theories covering different fields they merge into one; (c) or, less frequently, deeper understanding reveals two theories to be one. The first two changes are fairly clear; an example may help to clarify the latter. A wave (of sound, light, or any mechanical disturbance) can be represented in accordance with all verifying principles in two ways: (a) by means of wave train; (b) by means of the frequencies it involves and their amplitudes. One might speak of them as different interpretations of reality, both referring to the same sensory phenomenon. The latter reflects more immediately the harmony of a major chord, the former the characteristics of a single note. These two are related by Fourier analysis, which expresses the sound wave in terms of the amplitudes of its frequencies. Another more technical example of the fusion of two theories is encountered in statistical mechanics: (a) the older one is Gibbs' analysis of thermodynamics in terms of ensembles; (b) the other is a theory developed by Darwin and Fowler. Both explain known facts. The former is more easily applied to classical thermodynamics, the latter to quantum statistics. Both, however, explain the same experimental facts.

Complete reductionism is equivalent to a belief in the unlimited extensibility of a single theory. It often takes the naive form of supposing that a theory of simple phenomena (whatever that may mean!), when properly understood and refined, will ultimately explain all experience. And since physics is generally regarded as the simplest—or at least the most concrete—scientific subject (even though the modern physicist will probably deny this conventional belief), the most common form of reductionism is physicalism, the view that everything can ultimately be reduced to physical terms. The most naive form of physicalism is materialistic monism, the belief that everything consists of matter.

Such oversimplifications have generated a widespread terminology. Biologists and some philosophers are accustomed to speak of hierarchies and of levels, both terms referring sometimes to actual domains of existence (e.g., the inorganic, organic, and living world) and sometimes to theories explaining them. The word "hierarchy" literally denotes a pyramid of power, authority, or control. Its primary meaning is ecclesiastical control, for the word "hierarchy" is a compound of *hieros* (sacred) and *archein* (to rule). This usage has some relevance for biology (though more limited than is often believed), where specific parts of an organism monitor its actions and its genetic processes with ascending degrees of control. Evolution leads from primitive chaos to almost miraculous order, from organisms displaying simple behavior to complex and purposeful actions. It resembles a "temporal" bureaucracy, and the term "hierarchy" is applicable. Even "level" makes sense if it is applied to a temporal stage of the entire process.

Evolution is a subject most widely discussed in reductionist terms. Space will not permit its inclusion in the present volume, and we hope to give fuller attention to it later. In this context we ask the reader's permission for inserting a general, cautionary comment. For there is mounting evidence that old-style theories of evolution such as Darwinism, illuminated by Mendel's laws, and even Neo-Darwinism and the more recent astounding discoveries in the field of genetics—still lack essential explanatory evidence. Chance may have to be augmented by purpose. At any rate information and systems theory[5] are becoming better tools in this area than the theoretical techniques still widely used.

In the physical sciences the term "hierarchy" is rarely used, is in fact useless, and we shall dispense with it.[6] Our following occasional use of the term "level" requires comment. "Level" may refer to a certain degree or stage of complexity of existing things or aggregates of things, or it may denote more or less complex theories explaining them. It therefore is customary to distinguish between "levels of existence" and "levels of explanation." The former will here be discarded as unnecessary, for we have seen in previous chapters[7] that what is called physical reality is constructed by rules derived exclusively from elements involving human perceptions and reasoning. Hence, if the latter are more complex or more embracive, i.e., form a "higher level," so does the domain of existence that corresponds to them. The two theories therefore become synonymous.

These rather general remarks may be best understood by taking examples from parts of science that seem fairly complete and are most widely accepted. Most of them allow a description in terms of levels of explanation, though we shall later be forced to adopt a more specific language that is in accord with previous considerations and concentrates on what we called domains and realms of explanation, each with its own set of observables.

First, however, we must speak a bit about analysis. The scientific methodology we have outlined defines what we have called physical reality. But there is growing evidence, and an increasingly serious search, for modes of experience such as mystic states of consciousness, the attitudes reached through yoga and meditation, hypnotic awareness and dreams that are as human, and in their own ways as veridical as physical or sensory reality. These modes of experience have been aptly called alternate realities. They require an analysis of consciousness, perhaps in a manner not yet attempted.

To indicate what may be involved, let us mention only one difficulty. The external world, the physical reality, is established by a sequence of steps of which the first, called a rule of correspondence, or operational definition, translates conscious awareness (e.g., the feeling of hotness in my fingertip) into an objective construct, an observable (called temperature) that can be quantified. As a result of this crucial step various forms of mathematical techniques can be applied to the constructs thus defined, and numerically testable theories result. It is doubtful whether conscious states of alternate realities—or, indeed, normal feelings, moods, desires, volitions, etc.—can be similarly translated and quantified. We certainly do not at present possess even a language that can properly render the shades and tones of such experiences. A *faute de mieux* we use our sensory language metaphorically, employing expressions like those cited elsewhere. This clearly bespeaks a rather essential difficulty in any study of consciousness that follows the established physical pattern. It is not even certain that such important methodological principles as unlimited extensibility, the essence of our theme of reductionism, can be retained. Nevertheless, we shall assume it and see what results we may obtain.

1. We look first at some trivial examples because they exhibit very clearly some features of more complex ones. One-dimensional analysis of space is a most primitive kind of science. Its constructs are points and lines, and its only observables are distance and direction. The next "higher" level of geometry is represented by two-dimensional space, where we encounter points and line as well as polygons, circles, and other two-dimensional figures. Length and distance are still valid observables, but a new one "appears"—*area*. A one-dimensional being (a creature knowing only fore and aft and moving along a line) would have no direct conception of an area. That concept could not be visualized by it, nor would the creature introduce it into its world except possibly as an abstract conjecture. The reverse, however, is not true. A two-dimensional being can operate in a meaningful way with distances. If two-dimensional geometry is a sequential level (usually regarded as "higher")—say, of complexity, then we find what might be termed continuity of explanation in one direction but not in the other.

A change from the domain of two to three dimensions will introduce further observables, like *volume,* and new concepts like solid figures, with which solid geometry deals. Here again we witness continuity of explanation in one direction but not in the other, according to this terminology from above but not from below.

One-way continuity of this kind is typical of some less trivial examples as well. More important and more general, however, is the fact that the observables and the explanatory laws at the "higher" level (e.g., solid geometry) could not be visualized in terms of the "lower"-level observables. It could of course be conjectured as a mathematical tour de force.

2. We now turn to a physical example, the relation between Newtonian dynamics and thermodynamics, or statistical mechanics. Here we encounter again some of the features of our first examples, but in a less obvious way.

Newton's theory of motion describes the behavior of individual particles. Observables are mass, position, velocity, acceleration, and force. His three laws of motion regulate and predict these quantities for each individual particle and therefore define its motion.

Thermodynamics deals with large assemblies of molecules such as a stationary gas or a liquid, each molecule being subject to Newton's laws. But knowledge of all Newtonian observables is not obtainable because the individual molecules, whose number in the air of an ordinary room is of the order of a trillion trillions, can not be observed. Nor are they needed for the scientific description of the gas. In thermodynamics and its theoretical counterpart, statistical mechanics, new observables appear and are sufficient to deal with measurable properties of the gas: They are pressure, volume, temperature, and entropy. In terms of them we discover the laws of thermodynamics.

The relation between the observables of individual molecules—if they were known—and those of thermodynamics is that knowledge of the former at any given instant would allow us to calculate in principle (though not in actual fact because of the enormous difficulty of the calculation) the values of all thermodynamic observables. The reverse, however, is not true. Hence, Newtonian theory represents the "lower level." There is continuity of explanation from "below" but not from "above." And as in the first example the higher-level observables are meaningless at the lower level: A single molecule has no entropy and no temperature.[8]

Having "risen" from particle mechanics to thermodynamics we find that the further upward extensibility of the new theory is im-

mense. Temperature and entropy are not observables confined to finite fluids; their meaning stretches to infinite domains—in fact, to
the entire universe. The second law of thermodynamics implies that
the entropy of the universe continually increases. Nor are ideas like
entropy and temperature always material observables: We assign
"temperature" to the radiation filling empty space, "volume" to the
entire universe. Evidently the extent of the thermodynamic domain
is tremendous.

A word might also be said about the "reality" of the observables
associated with the two domains of our present example. Those encountered in Newtonian dynamics, like mass and velocity, are all
simple, direct, and easily visualized, as are the laws relating them.
Their relation to everyday experience is close. But the observables
and laws of thermodynamics are often abstract and defy visualiza-
tion. One of the most successful theories of thermodynamics was developed by Gibbs.[9] It operates with ideas like phase space, which has
six times as many dimensions as the gas has particles. It refers to an
"ensemble," which is a large number of postulated replicas of the gas
under discussion, each replica containing the individual molecules in
different Newtonian states. Here the philosophically minded reader
is prone to ask: Is phase space real? Does an ensemble exist? The answer, we feel, is affirmative, for those constructs play a necessary role
in our explanation of thermodynamic phenomena. But do phase
space ensembles "reduce" to the more elementary concepts of thermodynamics, do they "evolve" from them, or do they simply transcend them? We shall answer this question when we've accumulated
more information.

3. Another theory descriptive of the behavior of fluids is fluid dynamics, which deals with their motion. The reader might be inclined
to place it on a higher level than thermodynamics; but this, as we
shall see, is a disputable judgment. In its simplest form, to which we
shall limit our discussion, it ignores thermodynamic variables and introduces observables characteristic of fluid motion. Chief among
them are the fluid's "density" and its "current" (volume crossing unit
area per unit time). They are related by a law known as the equation
of continuity. The new observables are not derivable from those of
thermodynamics; they are additional to them. In this case there is no
reducibility, but the laws of thermodynamics and fluid dynamics can
be brought together and applied without conflict; they merge into a
larger theory; they are compatible.

4. When a fluid shows differences in temperature, a new phenom-

enon, "heat flow," occurs. Again we encounter new observables, which appear in a fundamental law called the heat flow equation. Some of these observables have no meaning for a steady state, are not reducible to its "level," and it is questionable whether they could have been foreseen. Again, while there is no reducibility in any normal sense, there is compatibility, a kind of fusion in which two theories merge into one.

5. Our third and fourth situations are among many in which the term "level" has become diffuse, where extension or range would probably be more meaningful. Electromagnetism, the next subject to be considered, presents an aspect that, in customary parlance, might introduce again a kind of reductive echelon, where one-way reducibility makes sense. As we shall see, electromagnetism involves a transition somewhat similar to the transition from Newtonian mechanics to thermodynamics or statistical mechanics, which provides the mathematical explanation of that transition. And as in that example the upper stage opens enormous vistas into further, incompletely explored realms.

The earliest and simplest description of electromagnetic phenomena involved the new observables "electric charge" and "magnetic pole strength" in addition to the universal concepts of time, distance, velocity, and acceleration. The basic law was Coulomb's, which says that unlike charges or poles attract each other with a force inversely proportional to the square of the distance between them and proportional to the product of the force-producing charges or pole strengths.

After Coulomb the theory was enriched by the addition of further observables like "electric" and "magnetic field strength," "Faraday's lines of force" and the concept of an "ether." At this stage no one could have foreseen the connection between moving charges and magnetic fields, which was discovered by the Danish physicist H. C. Oersted in the 19th century. Then our understanding reached the highest level of abstraction and complexity in the discovery of J. C. Maxwell's famous equations, which are relations between electric fields, magnetic fields, electric charges, and magnetic pole strengths. For details, see previous reference. These equations, which exhibit an unexpected but pleasing measure of mathematical elegance and symmetry, opened vast perspectives that were previously closed. The concept of the radiation field, which includes all forms of light, microwaves, X and gamma rays, finally evolved, by extrapolation and introduction of unforeseen observables, from Coulomb's law. Elaboration, if that is the correct term for the opposite of reduction, coupled with the creation and use of new observables transcending the earlier realms, opened expanses of physical reality which, like entropy and temperature, defied adherence to matter, filled space, and extended to infinity.

Our second example, the passage from Newton's laws to thermodynamics, and the present one, the passage from Coulomb's law to the higher form of electrodynamics, have these features in common. The observables needed in Maxwell's equation could not have been derived, discovered, or foreseen from Coulomb's law, even though Coulomb's law is logically implied by them. Yet if the observables entering Coulomb's law were known for every one of many charges producing an electromagnetic field, its observables could be predicted. The converse is not true: If a complicated field (electric and magnetic field strengths at every point of a region of space) were given, the location and velocities of the many charges producing it are not inferrable, for there are different charge distributions that can give rise to the same electromagnetic field—at least within a finite region of space. In both examples, though not in many others that could be cited, there is what in terms of levels might be called continuity of explanation from above but not from below. Both examples suggest the inadequacy of the ordinary idea of reduction and even some of its terminology.

The example of charges "producing" a field that extends throughout space presents another informative aspect involving the philosophic distinction between monism, dualism, and pluralism. Every charge, like every mass, contains energy. Energy is a real entity in physics as well as in the understanding of common sense, and one is prompted to ask where it is located. It is shown in textbooks that a charge q, residing on the surface of a sphere of radius r, has an energy q^2/r, this being the amount of work required to carry elements of charge in infinitesimal amounts from infinity, where no force acts on them, to the sphere which is its ultimate location, and thus presumably the place where its energy is concentrated. But the electric field that extends from the sphere to infinity also contains energy at every point, and the formula for its energy density (energy per unit volume) is well known. The quantity q^2/r results. Evidently we are dealing with the energy previously assigned to the charged sphere.

Where, then, does the energy reside? There is no reason for preferring one interpretation to the other. Still, the energy is a real
quantity.

Here is a situation where it is well to remember Alfred North
Whitehead's warning against the "fallacy of simple location." The
philosopher, confronting it, may be inclined to speak of a dualism,
reminiscent of Descartes' *res extensa* as distinct from *res cogitans*. In
the example here discussed physicists see no need to speak of a dualism; they regard the distinct aspects of energy possessed by a charge
as interesting but not perplexing; they may in fact be inclined to
think of the mind-brain problem, which will occupy our attention
later, in terms of the compatibility between the view of a charge sit-
uated on a body and that of an energy distributed through space.

We shall encounter the contrast between monism and dualism,
and likewise the mind-brain problem, on several later occasions. The
issue is central to the claims of reductionism and can hardly be set-
tled before our analysis of the latter is completed. But a word should
perhaps be said at this point about a new aspect of the monism vs.
dualism controversy in the face of modern physics. In the past the
word "body" had a simple, unique, and undifferentiated meaning; all
res extensae were of a kind. The same was true about *res cogitantes*.
The latter belief is still common, even though very little has been
learned about the nature of the mind, mainly for reasons that, as we
have already emphasized, forbid the application of physical methods
to mental states. But as regards matter, the body, an enormous
amount of new knowledge has revealed a degree of complexity, an
unexpected variety of onta, some of them fields of fundamentally different character, that is impressive.[10] The physicist is therefore likely
to ask: Is all matter really one? Does it not have components that are
essentially different, components that do not reduce to one another?
Most nuclear physicists would affirm such an essential difference,
would enlarge Descartes' distinction into one between *res cogitans*
and numerous *res extensae*. In other words, they would expouse a
pluralism instead of a dualism.

8 Reductionism, II

Let us review some of the relatively simple examples of reductionism we've discussed from a slightly different and formal point of view, to provide further evidence for the inadequacy of terms like level, hierarchy, and continuity of reduction, evidence that casts further doubt on the meaning the term "reductivism" (or "reductionism") claims to possess. We deem this elaboration important because in later portions of the book, where the questions of the reality of mental phenomena and their often attempted "reduction" to physiological, chemical, and physical processes is discussed, a full understanding of the implication of reducibility is important.

Consider once more the first example of the preceding chapter. We shall designate the domain of two-dimensional geometry by A, that of three-dimensional geometry by B. Correspondingly we speak of A-observables (line, area, shape, etc.) and B-observables (volume, surface, etc. in addition to the former). We then find that:

a) some of the B-observables mean nothing in A, but
b) the A-observables are meaningful in, and can be discerned from, B.

If, as would generally be supposed, the B-level is higher than the A-level, this state of affairs would be described by saying that the higher-level observables determine the lower ones: There is continuity of explanation from *above*.

In our second example the letter A shall designate Newton's theory of motion as it applies to individual molecules. Observables are mass, position, velocity, and force of each individual molecule. Level

B is the domain of a gas at rest enclosed in a vessel; observables are temperature, pressure,[1] and volume. Reductionists would speak of *B* as the higher level. Here we find that:

a) the *B*-observables mean nothing in *A*, and
b) the *B*-observables depend on, and can be "predicted" from, the *A*-observables

This state of affairs might be described by saying that the lower-level observables determine those of higher level:

There is continuity of explanation from *below*.

These two examples can be amplified by numerous others that show that continuity of explanation is not a one-way matter. Implied is a lack of meaning of the terms "above" and "below"—indeed, of the term "levels."

To give another physical example that is formally similar to the first—and that may, without much harm to what follows, be ignored by the reader who is not interested in detailed aspects of physics—we examine the relation between the pre-Maxwellian theory of electrodynamics, which operates with observables like magnitude, sign, position, velocity, acceleration, etc. of electric charges (we will call this "level" *A*), and the Maxwell-Lorentz theory of radiation, *B*. Here the *A*-observables are again meaningful in *B*, are in fact among the *B*-observables. But again, some of the *B*-observables are meaningless in *A*. We encounter what one might call partial continuity of explanation from *A* to *B*, complete continuity from *B* to *A*. The concept of level becomes confused. On the other hand, were we to regard *A* and *B* as identical, we should have no room for the radiation field, which belongs only to *B*. This example shows perhaps more clearly than the others that the concept of reduction must be replaced by transcendent elaboration with continuity without reference to levels and hierarchies.

It is also interesting for a special reason.[2] Having transferred attention from *A* to *B*, one concept—namely, the radiation field, where observables are electric and magnetic field strengths—takes on overwhelming importance, detaching itself, as it were, from its originators, the moving charges. First, because of its finite velocity, its present state depends on the previous condition of the electric charges (retarded potentials); it can even exist when the charges are annihilated. To complicate matters still further, the destruction of the charges (e.g., electron-positron annihilation) creates its own radiation field, which is superposed upon the former, then pervades all space in the absence of the material charges but retains its identity—though its observables will change their values in time.

A most important, cosmic illustration of this has occupied the attention of physicists and astronomers recently. The "big bang," presumed to be the origin of the present universe, must have occurred some 12 to 15 billion years ago. Precisely what happened could be inferred only from considerations of certain likely astrophysical processes, which included a cataclysm among onta, such as photons and charged particles, most probably the destruction of many of them. Whatever its details, whatever the precise nature and number of its no longer existing originators, the radiation field remained, has been identified and measured. The achievement has been honored by the award of Nobel prizes. A person cherishing a belief in immortality might be tempted to regard this as a physical example of the survival of a nonmaterial entity upon the death of matter. An elementary illustration of the same principles is the well-known fact that the light of a flashlight continues to travel through space even when the flashlight has been destroyed.

In the present context, however, these examples merely show the irrelevance of any kind of reductionism that does not involve transcendence, does not allow for the possible role of unpredictable entities in any given mode of explanation.

We now discuss a few instances of compatibility where the term "level" loses its appropriateness because the change in the observables encountered is so radical that, aside from the fact that those characterizing one "level" could not be "seen" or inferred from the other, as in previous examples, the "emerging" ones[3] defy both common sense and visual comprehension. They cease to be *"anschaulich"* and violate our knowledge of the molar[4] world. Two features characterize the related "levels": they exhibit not only different constructs and observables, as did the former, but require different *modes of explanation;* nevertheless, they retain their compatibility. The first marks the transition from the molar world to the extremely large, the second to the extremely small. Level then becomes a matter of size. High means large and low means small.

The following elaborate examples of transcendence with continuity are drawn from modern physics.

Kant convinced the scientists of the 19th century that time and space were what he called "transcendental conditions for the possibility of sensory experience," ideal forms imposed upon phenomena by the nature of the human mind and therefore "a priori," immutable, structuring all our sensations in specific ways. He believed that an analysis of time gives rise to arithmetic, that of space to geometry, and that the theorems of these two disciplines are unique. The structure he assigned to space is what we now call Euclidean.[5]

But toward the middle of the 19th century mathematicians began to wonder about the uniqueness of the theorems of geometry, and men like Bolyai, Lobatchevsky, and Riemann developed new types of geometry called non-Euclidean, which feature theorems differing from those previously believed to be universally true. However, because all scientific knowledge of the time obeyed the laws of Euclid, which Kant had claimed to be a priori, immutable, and universal, the new kinds of geometry were regarded only as interesting mathematical artifacts lacking physical reality, because they lacked contact with immediate experience. They were considered as internally consistent fairy tales. Space was infinite, parallel lines never met, and a straight line was the shortest distance between two points.

All this changed in 1916, when Einstein published his theory of general relativity. To be precise yet brief: He discovered that certain astronomical observations, puzzling to the astronomers versed in Euclidean geometry, yielded beautifully to the theorems of Riemannian geometry. This astonishing success elevated the latter from the status of a mathematical toy to a discipline describing reality. Its consequences included the acceptance of a new mode of explanation in which space was no longer infinite but had a finite radius; the shortest distance between two points in the neighborhood of a star was no longer a straight line; two bodies moving along parallel lines would meet after a long but finite time.

Einstein was fond of explaining curved three-dimensional space by asking his audience: Imagine a worm or caterpillar that has a sense of only one dimension. It knows only fore and aft and crawls along a line. Now let that line be a circle of large dimensions. The worm, crawling along it, would presumably feel itself proceeding along a straight line. Actually it is moving along a curve in two dimensions, but the curvature is conceivable only to beings that have a perception of two dimensions. Evidence of the curvature would be available to the animal only if it moved around the entire circle and came back to its starting point, which it recognized.

Now think of a two-dimensional creature like a flat bug that is capable of distinguishing fore and aft, left and right, but not up and down. It crawls on the surface of a sphere but can only think of it as a plane. We, capable of visualizing three dimensions, know that it is moving on a curved surface in three-space.

We now extend this reasoning. Think of three-dimensional beings moving in a space curved in four dimensions. They could not perceive that curvature and would believe they are living in a three-dimensional Euclidean space. To be sure, if they proceeded along what appeared to them as a straight line, traversing their entire-finite-space, they might return to their starting point. We are these three-dimensional beings, but our world is too large to traverse it. A light ray, however, sent forth along a great circle in four-space, would return to its starting point.

The question has at times been raised whether man's conception of space is intrinsically three-dimensional or whether he could learn to visualize the fourth dimension aside from conjecturing or constructing it. It is said that the mathematician Poincaré answered the question affirmatively, claiming he had trained himself to think in terms of four dimensions.

Later another incredible observation, hardly acceptable in the old interpretation of the universe, fell in line with the other astonishing proofs: The finite radius of space was continually expanding; the recession of distant stars and nebulae, known by the change in frequency of the light they emitted, was evidence of the expansion of space, of the increase of its radius. *Things utterly unthinkable in the molar world happened on the grand scale of the cosmos.*

As if to torment our molar intuition, time became the fourth dimension of space.

Had our concerns been restricted to our earth or to our solar system, the new observable—expansion of space and of the universe—would have made no sense, could not have been surmised. Yet the cosmic view, the new mode of explanation, is compatible with everything we know about our planetary system. To say that we have reducibility from "above" but not from "below" is still true, but it ignores the novelty, the strangeness of the new observables, which mathematicians call the curvature and the metric of space-time.

Some recently established consequences of the theory of relativity are truly fantastic. It leads to the prediction of so-called black holes,[6] very dense aggregations of stellar matter held together by gravitational forces, called black holes because they "swallow" everything near them, even the light that passes through their neighborhood.

Nor, of course, do they emit light of their own. Since they cannot be seen, their existence becomes manifest only by their carnivorous effects on nearby stars, and such effects are generally believed to have been discovered.

Strange things occur within black holes: the metric (geometry) of space-time is altered; time and space interchange their roles. To some theorists this suggests that conscious beings inside a black hole can go back and forth in time, reliving and preliving their life experiences. Aside from these speculations the following facts are to be noted.

A body held together by sufficiently strong gravitational forces possesses two kinds of energy: the well-known mass energy given by Einstein's formula $E = Mc^2$, E being energy, M mass, and c the velocity of light. In addition the body contains negative potential energy due to the attractive force of gravitation holding the mass together. For a given set of values of the mass M and the radius R of the body (assumed to be spherical) the negative term cancels the positive one. This means the body has zero energy (cf. Jordan[7] and Open Vistas[8]): its emerging out of empty space would contradict no law of nature. *Creatio ex nihilo,* St. Thomas' belief, becomes a scientific possibility. What we would surely call a miracle is sanctioned by science.

Two further points, apparently not widely known, are equally astonishing, indeed mystifying. The mathematical condition for a black hole (known as the Schwarzschild condition) is very similar to the zero-energy condition, making it probable that black holes can spontaneously spring into existence. And finally it turns out that within the uncertainty of our knowledge of the total mass and the radius of the universe, these parameters may possibly satisfy, at least approximately, both the conditions for a black hole and for zero energy.

Hence one might raise the questions: Do we live in a black hole? Is the total energy of the universe zero? Common sense will say no, and we feel that more accurate data (ours were taken from Harlow Shapley's work of some twenty years ago) will bear out the verdict of common sense. We mention the matter as one of the instances where recent theories of physical science bring us to the frontier of what seems at present to be inconceivable.

These results are daring and highly conjectural. But they are compatible with the mode of explanation prevailing in contemporary astronomy, yet wholly unforeseeable from the "level" of that science a few decades ago. We forego a recital of the new observables that have appeared, observables as foreign to the Kantian conception of time and space as are entropy and temperature to the Newtonian mechanics of particles.

The preceding example marked our passage from the molar world to the macrocosm. We now take the opposite course into the microcosm, the world of molecules, atoms, electrons, and other so-called elementary particles or onta. There, too, we find that ordinary observables lose their relevance, their usefulness, and even their *meaning*. The philosophic difference between the observables in the two domains—or, *better*, in the two modes of explanation—is even greater than in the former instance.

The most important axiom, even though it is philosophically trivial and obvious, to remember in our passage to the microcosm is that *entities too small to be seen cannot be endowed with visual attributes*. It should therefore not surprise us that molecules and atoms, and to an even greater extent their constituents, acquire characteristics that seem strange to common sense. These characteristics are indeed observables although direct observation of microcosmic events is impossible.[9] To avoid further complication of language we shall continue to use the term "observable," defining it as any quantitative construct related to an observable, sensory event by an operational definition or, more generally, a rule of correspondence as defined in chapter 3. In this sense, then, the mass of an atom or the charge of an electron remains an observable. Their conception does not tax our imagination. But there are others that prove more troublesome.

Let us consider briefly the hydrogen atom. The last attempt to represent it in visualizable forms was Bohr's, whose theory is well known and is still taught in elementary physics courses. According to it the atom consists of a central core, a proton, and an electron that revolves about it with enormous speed. The distance between them is about 10^{-8} cm (one-one-hundred-millionth of a centimeter), and the sizes of proton and electron are, respectively, about 10^{-13} cm and 10^{-10} cm, the speed of revolution about 10^8 cm/sec. This presents a pleasing picture. When magnified by the factor $10^{13} = 10$ trillion, the proton takes on the size of a marble, the revolving electron the size of a 30-foot balloon, and the balloon moves about the marble approximately 10^{15} times per second at a distance of nearly half a mile.

We shall try to show, by way of a simple consideration, that this picture is inconsistent. A revolving electron has a continuous path and therefore a definite position at every instant. Suppose we try to determine, to measure that position. About the only way this could be done is to try to hit the revolving electron with another one, say another electron or a photon. Intuitively, the direction of motion of the reflected projectile could tell us the place at which the orbiting balloon was located at the instant of contact. An experiment of this sort is in principle (though hardly in practice) feasible. So assume it gave us an answer leading us to suppose that the balloon was at the top of its orbit.

Unfortunately this result makes our entire measurement meaningless. For the same theory (electromagnetism) Bohr used to develop his elegant model implies that the reflection of the projectile, e.g., a photon, cannot be instantaneous but requires an interaction that lasts a finite time, in this instance something like 10^{-12} seconds. But since it revolves 10^{15} times per second, it went around the proton about 1,000 times during the interaction. Hence to suppose that it was located at one point of its path is meaningless. This paradoxical result, the impossibility of measuring the position of our electron, and any other entity of comparable mass, is characteristic of *all* efforts to measure, to determine empirically, the entity's location when it moves with a definite velocity.

Should we then say that the electron does not have a position, does not pursue a continuous path, or that position is no longer a proper observable? Before answering this question hastily, we note a complicating fact: There are circumstances under which an electron's position *can* be measured. If a beam of electrons moves toward a screen that scintillates when one of them impinges, then the scintillating spot does mark its location at the moment of impact. The difficulties arising in connection with the hydrogen atom do not appear. Evidently there are conditions, called *states*, in which the electron manifests a true position, and others in which it does not.

What has just been said about position is true for all other classical observables, such as velocity, momentum, and energy.

An adequate account of this anomalous situation is given by the so-called *latency theory* of observables.[10] In classical molar physics a definite value of an observable—let us say energy, for example—is present in every possible state of a given system. The system *has* that amount of energy. If forces act on it, the value may change in time, but it *has* one value at every instant, and if the state is known in its temporal variation, the value can be predicted.

In quantum mechanics, the theory describing the microcosm, such is not the case. Crucial at this point becomes the meaning of the word "state." In "classical" (prequantum) physics the state of a system is a set of observables, a definite state a set of *values* of observables. The state of a particle is defined by the values of its position and its velocity, the state of a fluid by the values of its pressure, volume, temperature, etc. What we have said above indicates that, since observables in quantum mechanics may not have values in this simple sense, a new definition of "state" becomes necessary. We shall give it the symbol φ without specifying at this point its precise significance (we will deal with this issue presently). Thus, without saying on what it depends and how it is determined, we first comment on its relation to observables.

Let φ denote the state of an electron. Among its observables are position x, velocity v, and energy E, although we have just seen that they cannot always be determined as truly possessed attributes of the system. A typical state φ may be one in which the energy has a definite value, but x and v do not. This means that whenever the state is present (and there are known ways of preparing it), E will yield upon measurement the definite value E_1. But if x or v is measured, a great variety of different values may result. In that case we call E a *possessed* observable when the electron is in state φ, and we call x and v *latent*. They are not really present but somehow become realized in the act of measurement, resulting, one may suppose, from interaction with the measuring device. Latent observables yield different values upon repeated observation; possessed ones do not. The philosophy surrounding the concept of latency need not concern us here. Heisenberg, who endorsed the idea, preferred to call it potency[11]; a term he owed to Aristotle.

The special state φ in which E is a possessed observable is called an eigenstate of E, and if the value of E that turns up in every measurement is E, we call it φ_i. For every known observable there exists eigenstates, but there are rules forbidding the occurrence of eigenstates in which certain pairs of observables are possessed. Thus the uncertainty principle demands that in any state φ in which momentum is possessed, its position must be latent, and vice versa. There are other pairs of observables that cannot both be possessed.

In addition to the idea of latency, quantum mechanics features what its name implies—quantization. This is to say that certain observables, notably E, even when possessed, can only yield one of a specific set of values, say E_1 or E_2, . . . or E_i . . .[12] These are said to be the eigenvalues, and the state in which E_i is possessed is called eigenstate, φ_i.

The third important feature of quantum mechanics is its replacement of old-style causality, i.e., mechanistic determinism, by disciplined and measurable chance. Suppose that a physical system such as an electron is in a state in which an observable like E is latent. Hence, when a measurement is made, a certain value E_i will turn up. Aside from the fact that it must be an eigenvalue, one of a known set, we cannot predict what it will be. However, when φ is known, the probability with which any one of the various E_i will appear can be calculated. More specifically, every time the state φ is prepared and E is measured, a different E_i will generally appear, but we can predict how many times in a great number of such sequences of preparation and measurement a given E_i will be found. We know the relative frequency of the occurrence of every E_i. If $\varphi = \varphi_i$, E_i is possessed and the relative frequency, also called probability of measuring E_i, is 1.

We might say that the concept of latency has a probabilistic structure. For a given φ, in which x is latent, the electron has no position. To speak of it as performing a motion in the accustomed sense would be improper. Indeed the claim that it behaves like the angels of St. Thomas, who could proceed from one place to another without traversing the intervening distance, could hardly be disproved. But there is structure within latency, namely the predetermined probability distribution for the outcome of all possible measurements of all relevant observables.

We conclude our survey of the explanatory modes allowing us to understand the microcosm by reiterating that the very term "observable" has undergone a drastic and totally unexpected meaning. Since an observable can be latent, a specific set of observables like x, v, or E (position, velocity, and energy) can no longer be employed to define a state in general. We face a new state definition, φ, the state function. And if we know φ, we also know that some observables are latent, yet we know the probability with which they occur. The structure of reality has been shifted from old-style observables that were more or less *anschaulich*—intuitable is perhaps a satisfactory translation—to φ-functions, from certain knowledge of what will happen to probabilities.

If this sounds strange or even deplorable, let us recognize that probabilities, too, are observables. They satisfy all the methodological requirements to be imposed upon physical observables. In particular, they are both measurable, hence quantitatively significant, and represent a construct enjoying extreme extensibility and entering into fertile relations with innumerable other useful constructs.

Hence we do not hesitate to assert that probabilities are part of physical reality. We shall return to this aspect of reality when we later
discuss the problem of free will.

One final point. Are all observables latent? More precisely, can
every observable appear in a latent role, or are there some that are
always possessed? The answer to the last question is at present affirmative. Two of the old-style observables, charge and probably
mass,[13] are always to be possessed no matter what φ is considered.
This almost seems like an anomaly, leading one to suppose that our
theory is not yet complete.

As we now look down or up from the perceptible world that surrounds us, do we see a discontinuity in our mode of explanation? The
differences are so great that one is tempted to answer yes, but that
answer is false. For there is one feature, early recognized by Bohr—
in a limited version, to be sure, and often called the correspondence
principle—that we have not yet mentioned. It is this. The laws of
quantum mechanics are such that, as sizes and masses increase, latency gradually disappears and probabilities shrink to the values 1 or
0, i.e., become certainties. And this is no post-hoc addition to the theory but inheres in, and is demonstrated from, the complex but elegant mathematical foundations of quantum mechanics. Thus, while
the constraints and the observables of microscopic theory could not
have been "seen," or foreseen, from the "upper level," while they do
not even make sense when applied to the molar world, compatibility
exists.

Another anomaly can be seen in this example. It offers modes of
explanation that are what the exponent of reductionism would have
to call *reciprocally* reducible, where in fact no reduction takes place,
where modes of explanation remain on the same level. A case in
point is Schrödinger's quantum theory, which works with observables like φ, the state function, and observables like energy, which,
being latent, are represented as mathematical operators. Heisenberg,
on the other hand, associated the concept of energy and other observables with matrices, a state like φ, with a vector (usually in a
space of an infinite number of dimensions). For a while the existence
of two valid systems of explanation of the same set of phenomena
seemed amazing to physicists. But soon the originators of the quantum theory themselves proved the mathematical equivalence of the
two theories.

Here we are confronting a situation again quite different from
what conventional implications of reductionism would lead us to ex-

pect. Perhaps it should be called biperspectivism.[14] And when von Neumann invented even a third approach to quantum mechanics, did he introduce triperspectivism?

Viewing now our sixth and seventh examples in retrospect, the reductionist would doubtless have to reason as follows: The two examples deal with the microcosm and the macrocosm. Between them lies the molar world, which is amenable to treatment by classical[15] mechanics (aside from other related and compatible theories). Consistency would compel the reductionist to acknowledge three levels: The lowest (1) would be the microcosm; next higher (2) would be the molar world; and the highest (3) would be the relativistic macrocosm. And he would have to say that level 1 reduces upward to level 2, while level 3 reduces downward to level 2. Little is left of his hierarchy.

Because of the perplexing strangeness, the abstractness of these results we insert here a brief section that might make the continuous transition from the microcosm to our molar world a little more vivid and comprehensible.

9 The Worlds of Einstein and Heisenberg

IT IS SOMETIMES AMUSING—and usually shocking to the person who has fixed ideas about ultimate reality—to speculate on how the world would look to a frog, an insect, a fish, or to a being whose eyes are sensitive to X rays rather than light of the visible spectrum.[1] We hope you will not take it amiss if we indulge in the use of a similar approach to illuminate the nature of the microcosm. We shall invent a conscious observer of atomic size, whose visual organs have the sensitivity and selectivity of the most delicate conceivable physical measuring devices, whose time sense is greatly dilated to the extent of being able to see typical atomic events whose duration is of the order of 10^{-8} to 10^{-15} seconds (one hundred millionth of a second to one quadrillionth of a second). He can "see" individual electrons and photons and "observe" their behavior among a finite, discernible collection of atoms constituting a part of a material substance.

As observers we would perceive no coherent objects, at least not in our immediate vicinity. We would be aware of single darts of light (photons) cast off by single, spontaneously luminous atoms. Our world, the microcosm, is not uniformly illuminated and filled with moving things; it presents a speckled kind of vision with bright patches emerging here and there from utter darkness, different patches having different durations. Distant atoms, perceived as larger groups, do exhibit a kind of uniform glow and manifest a measure of cohesion in this chaotic milieu, but the small scintillations nearby give very little indication of uniformity or pattern.

If our observers are positivists or avowed empiricists who lack imagination and insist on constructing their world out of immediate sensations, they would hardly believe in the existence of permanent bodies at all times; indeed, they might doubt the existence of individual entities except at the instant of perception. They might even find it implausible to speak of the "flow of time," preferring the phrase "emergence of sensed intervals." Continuous time, continuous space, would probably seem inadequate expressions; discreteness, quantization would describe their experiences more aptly.

Our microscopic observer will see things under two conditions only: They can illuminate them by means of an external light source, or, in the case of an atom, they may wait for it to emit a photon. The manifestation of the presence of microcosmic objects will in either case be random so far as individual events are concerned. Randomness is of course an important characteristic of many events in our macrocosmic world, and it might be useful to reflect for a moment on how we, as macroscopic observers, pass from perceiving the randomness of individual events to perceiving regularity and continuity.

To this end, think of the motion of a firefly in a dark summer night. We see it emitting light at different points of space, yet we associate the discrete luminous points with a continuous path. Why are we not satisfied with the aesthetically charming element of randomness and abandon the making of a path? The answer to this question is partly metaphysical, partly empirical, i.e. there are two path criteria. The metaphysical reason for assuming continuity is this: If the luminous points are plotted, they lie on a smooth curve. This alone conveys a very strong conviction, for among the guiding principles by which we establish physical reality are simplicity and elegance of the concepts it involves, and in these regards a smooth curve, continuously pursued, has clear advantages over sporadic, discontinuous emergences. The second reason is obvious and clinching: We can see the firefly in the daytime and document its continuous path by direct observation.

Now consider the electron, which, according to an early view, describes a circle about the proton in the normal state of the hydrogen atom. Our microscopic observer will not testify to this allegation. He will see sporadic appearances in different places when the electron reflects a suitable photon—i.e., whenever its position is determined. These positions are discrete, like the luminous emissions of the firefly. Applying the two path criteria we mentioned we find, first, that a line drawn between successive appearances of the electron will not be a smooth curve but a highly irregular zig-zag graph, the corners of which (i.e., the presumable positions of the electron), will lie within a vaguely defined ring of finite width about the proton. And as for empirical verification, there is no daytime in which the electron can be seen.

The only evident regularity upon which theory can seize is the ring of dots, which marks the electron positions. Closer examination shows that these dots are densest along a circle, which, perhaps surprisingly, is identical with the path known as the Bohr orbit, the circle along which, according to the early Bohr theory, the electron should have moved. But it also appeared elsewhere, for the dots occur at points away from the circle, though with decreasing frequency as the distance from the circle increases.

The simplest interpretation of these facts, and the one espoused by the quantum theory, is that the density of dots within the ring represents the *probability* of the electron's position. We cannot tell where the electron will be at a given instant, but we know the probability with which it will appear there. The ordinary observable, position, is not subject to regular laws, but laws which merely control the probability of various positions.

In this single instance we have exposed the major theme of the quantum theory. For what has been discussed with respect to position is true for most other observables in the microcosm: Measurement does not necessarily reveal consistent values of velocity, momentum, energy, etc., of our micro-entities, but it does yield probabilities of finding specific values of these observables. More specifically, if we ask what is the velocity of an atomic particle at the present time, we usually get an erratic answer. But if we ask: How many times in a million observations, made when the particle is in a well-defined, preparable state such as the ground state of the H-atom, will its velocity have a specified value? nature will give us a definite answer. What the observer of the microcosm has learned is that *probabilities become true observables,* whereas the customary ones are capricious.

We now ask our micro-observer to look away from his single atom and survey what happens in the distance, where many entities, atoms, and electrons are seen at once. There he will note some degree of coordination. The *mean* position of what he construes as objects seems to obey definite laws. If a list were made of the appearances in space of a group of luminous dots, and if their mean positions were computed at every instant in the manner in which one finds the point called the center of population, this center would move more or less in accordance with macroscopic laws. In fact, nature relieves us of the need for this computation by performing it herself. We have already noted that the more distant objects of the microcosm show some coherence and consistency to our Lilliputian friend. This is because they are made up of many entities and comprise large masses of which only the mean position can be observed. The erratic behavior of individuals is thus obscured, and we confront relative certainties. And finally, perhaps amazingly, we recognize these regularities, these relative certainties, as the laws of nature we had previously discovered in the molar world.

After our imaginary excursion into the microcosm, which strained the resources of common sense, let us now allow our thoughts (always restricted by the necessary accord with valid theory) to travel into the macrocosm.

We choose a rather extreme example of space travel. A scientist given to adventure wishes to visit a star 1,000 light years distant from the earth, intending to return within a lifetime, let us say in forty years. Although this seems impossible because light would require 1,000 years to reach the star, it is nevertheless possible in theory for him to accomplish his mission. If he travels at a speed very nearly equal to that of light (186,000 miles per second) indeed, only 3.2 miles per second less than the speed of light, he will reach the star after twenty years. The reason is that his clock or watch, if it keeps going, suffers an enormous retardation and we shall suppose—as physicists invariably do—that his life rhythms as judged by an observer on earth suffer an equal retardation.

During his trip he will fly past other stars, all of which appear as thin disks with their axes in the direction of his motion. In the neighborhood of a star the direction of the flight would be slightly altered because of the star's attraction, and we shall suppose that our astronaut, in choosing his original flight direction, has made allowance for all such alterations. Above all, of course, he must stay far from black holes.

The problem of how our astronaut will manage to land or, in case he does not land, how he will turn around, will be left as an exercise to the reader. His energy of motion is equivalent to that of many H bombs; hence only some cosmic process, which he would hardly survive, would reverse his motion. But in theory it is possible, and he would return to earth after forty years. The earth, however, will be 2,000 years older.

What our ultrafast friend has learned is this: In the realm of the macrocosm time and space cannot be separated; mass length, and velocity, observables that seemed relatively independent in the sensory realm, enter into unique relationships.

The speed of rockets now available is but a very small fraction of that necessary to perform the task described. But experimental evidence has been sufficient to test, in an elementary way, the theory upon which the above example was based. To strain the ordinary notion of reality still further, we mention the well-founded hypothesis of anti-matter. It appears from the theory of relativity that, for every *on* there exists an anti-*on*. Thus, there should be bodies composed of onta as well as bodies consisting of anti-onta. Electrons and positrons are the simplest examples of opposite onta. Others have been observed in nuclear interactions, but as yet no one has seen an entire atom and its antiatom. Nevertheless, a principle of symmetry advocated by many physicists requires the existence of antiatoms and probably even of bodies composed of them. Since they are not found in our neighborhood, it is reasonable to suppose that they are elsewhere in the universe, perhaps in the form of antistars of even antigalaxies. This hypothesis gains plausibility on the following grounds. Matter attracts matter in accordance with Newton's inverse square law. A law of the same form, though of different strength, is Coulomb's law, which expresses the force between charged objects. That force, however, is attractive for unlike charges, repulsive for charges of like sign. The law of gravitation for ordinary matter is always attractive: Perhaps antimatter completes this story and provides the missing repulsion. This leads to two possible conjectures which provide the missing link. Either antimatter repels antimatter or it repels matter. In the first case antimatter will have been dispersed throughout the matter universe. In the second it will linger somewhere in space far away from matter.

When matter meets antimatter, both will be annihilated and converted into energy. For bodies of visible size this would produce an explosion of magnitude far greater than that of a hydrogen bomb. If an antistar were to collide with a star, the explosion would be of enormous astronomic force. So far as is known such a phenomenon has not been observed, but its possibility, strange to common sense, to the classical picture of reality, cannot be rejected.

10 Reductionism, III: Transcendence with Compatibility

BEFORE ENTERING THE DISCUSSION of other problems and disciplines, for which most common explanations often make use of rather simple, sometimes crude forms of reductionism, we summarize what we have learned from our consideration of the disciplines that are presumed to be best understood. Our examples have shown the term "hierarchy" to be generally meaningless: The concept does not characterize physical theories. Difficulties attending its use arise perhaps primarily from an ambiguity, a duality inherent in its interpretation. If the distinction leading to the concepts "high" and "low" is made on the basis of size or complexity of the *objects* to which a theory refers, then classical physics, the molar realm, lies above the microcosm. If it is based on the complexity of the *theory*, the opposite is true, for the laws of quantum mechanics are far more difficult to grasp than those of classical physics.

Obviously, the same difficulties attach to the word "level," whose use is standard in previous discussions of reductionism. The only sciences for which the terms "hierarchy," "levels," or "organization" are meaningful are descriptive botany and zoology. In general science is not a bureaucratic system, not a hierarchy of levels. Even such terms as "higher" and "lower," "inner" and "outer" may also be meaningless.

Henceforth we shall mean by "level" a *mode* or *domain of explanation*, and by "reduction" the reversal of *transcendent but compatible elaboration*, a concept we will now explore.

This drastic change may seem unfair without a brief reference to an older work (published in 1909) in which many experts have stated their views. The book is *Hierarchical Structures* by White, Wilson and Wilson.[1] Like most other treatises dealing extensively with the problem of reducibility of theories, it is a collection of essays by many authors, contains revealing insights but no single view, no synthesis of the meaning of its title, not even a consistent terminology. The distinction between hierarchy and level, made by Mario Bunge, is ignored by other authors, and nowhere does it become entirely clear that the philosophic issues under discussion hinge solely on modes of explanation and their inclusiveness, on the coherence or lack of coherence of theories, and above all on the relation between different sets of observables employed in different explanatory modes or realms, or simply theories.

In these deliberations we have profited from the analysis of Bunge,[2] who has subjected reductionism to logical as well as epistemological scrutiny. He, too, criticizes the unprincipled use of the term "level" but retains it in the limited region where it does in fact apply: in a set of domains characterized by the *size of their constituents*—i.e., the levels, recognized chiefly by biologists, which can be arranged in the rather obvious sequence: cell level, organ level, organismic level, population level, ecosystem level, biosphere level.

We are interested not so much in size of constituents as in the theories that regulate their behavior, in the observables called into being as we pass from one domain to another, in their complexity and their coherence. What matters in our approach is the interconnection between fundamental theories or, as we have occasionally expressed them, between modes of elaboration, between old and new observables that appear in them. Hence there is this difference between Bunge's (entirely correct) analysis and ours: He emphasizes parts and sizes, while we consider the nature of observables and the nexus of their interrelation, and all this from the philosophic views of the Nature of Physical Reality,[3] which defines physical or sensory reality, and Alternate Realities,[4] which gives evidence that epistemological approaches should be sought that explain states of consciousness other than the standard one.

The appearance of new observables Bunge chooses to call "emergence," a term whose use he shares with others, among them Karl Popper,[5] though not always with precisely the same meaning. For reasons already stated, we replace it by the stonger word, "transcendence." Strictly speaking, what emerges was already there, invisible and unexpected. We, however, wish to emphasize the uniqueness of the new observables, their creation by a new theoretical approach, the scientist's inability of simply conceiving them from a domain of explanation in which they have no meaning. Thus, the observables of three-dimensional geometry do not emerge through any kind of mental manipulation of distances and areas in two-dimensional space; they simply transcend it, even though the formulas for volume (which is inconceivable in two-space) contain areas and distances. And the same is true for the meaning of temperature, which cannot emerge from the laws of Newton, but simply transcends them. In this instance it is also worth noting that, when the theory of statistical mechanics, the basic explanation of thermodynamics, was developed, thermodynamics did in no sense "reduce" to ordinary dynamics, nor the reverse; the theories simply fused together into a single one. This happened because new concepts like entropy, free energy, which transcended mechanics, had been discovered and incorporated in the larger system of constructs.

In closing this discussion it should prove helpful for us to state our conclusions very simply in terms of the diagrams employed in the earlier chapters. We are indebted for the material that follows to Professor Harold Morowitz, who felt that a pictorial representation of our conclusions would be useful.

The meaning of reduction, which we have replaced by the term "transcendent elaboration with continuity," is most easily explained by reference to our P-plane–C-field figure. For what it implies is simply an advance to the left in the C-field, a further departure from P, usually coupled with an extension of the range of constructs parallel to the P-plane. In this sense transcendence means nothing more than systematic increase in understanding of the world. We choose as our example the theory of motion. In the time of Parmenides motion was explained as an illusion: It did not exist. Another Greek philosopher, Leucippus, who first introduced the atom as a constituent of all matter, went to the opposite extreme and declared that motion was an innate, external property of all atoms. The conception of rest became an illusion. These early conjectures do not conform to the method of science as we have outlined it and are reintroduced here are prescientific approaches to the problems of dynamics.

A theory vaguely satisfying the rudiments of our scientific epistemology was Aristotle's. He accepted the view that all matter consisted of four elements: earth, water, air, and fire. The natural place of fire was in the heavens; water occurred mainly on earth; air was above it. Each body's natural tendency was to remain at rest or move to its natural place; hence on earth fire normally rises, water and stones fall, air will rise in water and in earth to reach its natural places. These motions were called natural motions.

But a stone could also be thrown upward. Such motion was called violent by Aristotle and required a force. The stone would perform a violent upward followed by a natural downward motion.

Were we to represent this "theory" in terms of the diagrams of chapter 3, it would form one of the innermost contours of Figure 7. Immediate experiences P_1 and P_2 would lead by reification and obvious operational definitions to the observables h, heavier than its surroundings, and l, lighter than its surroundings. These observables are related by a law to motion, M, which (although the double lines were omitted in the figure) is also related to P. The observables, defined by crude rules of correspondence that can hardly be quantified, are motion, weight (h and l), and force, the latter crudely defined as push or pull. Aristotle's primitive theory of motion is roughly represented by contour 1 in Figure 7.

In modern physics we observe masses, positions of bodies, velocities, and acceleration. Arising in P, they take us via rules of correspondence to systems, bodies, or onta into C. The first law that satisfactorily combined some of these observables—i.e., these constructs—was Galileo's: All bodies fall with an acceleration of about thirty-two feet per second per second (contour 2). It was of limited range and dealt only with bodies falling near the earth's surface.

Galileo's law was refined and its range of application was greatly extended when Newton announced his three laws of motion and gave a new and more precise definition of the concept force (F instead of the older f; the two are but vaguely related). This important innovation together with Newton's law of universal gravitation made possible an understanding of Kepler's laws of planetary motion, which had produced a "scientific revolution" and thereby achieved an incorporation of a part of astronomy into the domain of physics. Furthermore it explained Aristotle's thesis; hence contour 3, meant to depict Newton's theory, includes both contours 1 and 2.

While the centuries following Kepler and Newton brought further elaboration and refinements to the theory of dynamics, no essential new elements were introduced until Einstein advanced his theory of universal gravitation also called the theory of general relativity. It

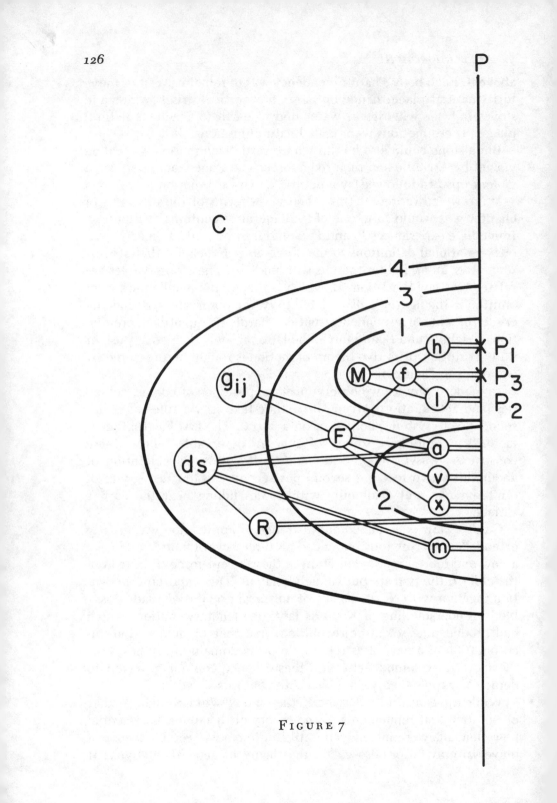

FIGURE 7

explained several previously mysterious *P*-experiences (the bending of light rays near heavy conglomerations of matter, the precession of the perihelion of the planet Mercury, a frequency change in light emitted by a heavy star) by introducing the concept of non-Euclidean space, which is characterized by the constructs g_{ij}, ds, and R in our figure, along with several others related to it.[6]

The more abstract (mathematical) a theory, the farther it extends to the left in the *C*-field. If its contour marks the present boundary of knowledge, it acts as a postulate beyond which science turns into hypotheses and conjectures.

What is the connection of our diagram with the problem of reductionism? Our analysis has led to the conclusion that reductionism is simply the passage from right to left in our diagram. In breaking through an established contour it introduces new, transcendent observables that may or may not be meaningful in the old domain. Associated with the movement to the left is often the unification of two previously separate domains, like 1 and 2 in Figure 7. All of this is what we term "transcendence with continuity."

This, however, is not the only meaning of the word "reduction." Strangely, a physicist may say that Einstein's theory *reduces* under certain conditions to Newton's. This implies a movement to the right in our diagram and hence a reversal of the primary meaning of reduction. The condition for Euclidicity is: All masses must be small, which makes space. In more precise parlance, the latter "reduction" is the progressive approximation of the predictions of one theory to those of a simpler one under given conditions.

Throughout this chapter we have discussed and exemplified the significance of transcendence. Now, a word must be said about the meaning of the accompanying adjective, "compatible." We understand it first of all in its simple, logical sense, where it implies freedom from contradiction; obviously, if the consequences of one theory contradict those of another, one of the theories will be rejected. But our use of the word assigns to it two other more specific meanings.

One characterizes predominantly mathematical theories such as geometry, where it might be expressed by the adjectives complementary or even supplementary. Consider the relation between plane and solid geometry. The latter makes use of the former: a cube is described as an object or a volume having six surfaces of equal area, each surface being a square. Observables in one domain are needed in the other. A similar relation will be seen to exist between algebra and arithmetic, between Riemannian and ordinary Euclidean space, and so on.

The second meaning our use assigns to the word "compatible"—and perhaps the most important to later sections of this book—appears in all domains of physical science. Here it suggests the possibility or indeed the occurrence of an *interaction*. To cite simple examples: The thermodynamic observable of temperature affects the specific gravity of a body; thermodynamics impinges on mechanics. The mechanical action of a thermostat causes a motion that closes a switch and allows a current to flow. Conversely, an electric field exerts a force on a charged body.

More elaborate examples connect the microcosm with the molar world, with what we called the see-touch realm. Along with many other onta an electron cannot be seen. Nevertheless it has effects in the see-touch world, for it leaves tracks in a cloud chamber that are visible. A single photon produces a scintillation on a screen; it can even have a biological effect in producing the mutation of a gene. Similar interactions occur between the macrocosmic domain, which obeys the laws of general relativity, and ordinary optics. The recession of nebulae is detected in the shift of spectral lines, by their displacement from normal positions on a spectrographic plate.

As we've just said, it is this third meaning of the word "compatible" that will demand our attention in some of the chapters that follow. To open your view we record two instances of interaction in advance. An unaccustomed (and often rejected) observable called purpose, we shall see, interacts not only with the biological-physical process of evolution by putting a bias on chance, but with many other domains of life as well. In connection with the mind-brain problem we would point out that the mind, though transcending the body, nevertheless interacts with it in demonstrable ways.

How far to the left in our diagram the penetration of the C-field may go is unpredictable. Will it encounter such concepts as consciousness and mind? Indulging in speculation at this point we might expect it, but doubtless by a process of transcendence rather than material reduction. It should thus be viewed as a theoretical system not wholly explicable in terms of, but yet interacting with, constructs of the physical realm. And in further elaboration of our theory we hope to show that the two lateral sides of our diagram may be joined, as though the picture were placed around a vertical cylinder. The mind, the ego, may then appear at the far right of the P-plane, where it merges with its constructional counterpart. The intervening space between the *P*-plane and the ego, the space at present empty but containing the observables of the mind and their relationships, would be filled when an explicit theory of consciousness is available. But these are conjectures about future science.

11 Causality, Epistemic Feedback, Purpose

O F THE GUIDING PRINCIPLES we have mentioned, causality has been and continues to be the subject of most philosophical as well as scientific interest and debate.[1] In Western science the first statement of the causal principle appears in the philosophy of Democritus, who said: "By necessity are foreordained all things that were and are and are to come." Aristotle's treatment of the topic was more elaborate; he made the distinction between formal, material, efficient, and final cause. The formal cause is the idea to be realized in a causal process; the material cause is the substance undergoing a change; the efficient cause is the external compulsion, the motive force; the final cause is the goal to be reached. When a sculptor makes a statue, the material cause of the statue is the stone from which it is made; the formal cause the idea, the quality of the statue; the efficient cause the exertion of the sculptor; the final cause the purpose it serves, its beauty. The first two of Aristotle's causes have dropped out of consideration, and of the latter two the final cause has become purpose, and the efficient cause is the one that has survived and remains a matter of scientific concern.[2]

As to the precise nature of causes, Aristotle's conception was rather indiscriminate; he drew three distinctions: things (the acorn is the cause of the oak), events (an accident is the cause of death), and states of physical systems (the present position and velocity of a body determine its future position and velocity). Increased refinement of the causal concept has rejected things as causes and takes events to be partial causes. The past state (known in terms of all significant observables) of a physical system implies, via laws of nature, the state at any future time. But what is meant by state?

Strictly speaking it is the set of all observables in terms of which a prediction of a future state can be made. We hasten to illustrate this heavy statement by considering a simple example. Take an ordinary object like a stone. Among its numerous observables are its mass, size, shape, color, position, and velocity in space. Each of these is important in some limited area of the molar realm; its size and shape interest the builder, for example, and its color the artist and the optician. We select for causal consideration the motion of the stone. Here it turns out that among all observables only the following are important: its mass, present position and velocity, and the force acting on it. The law relating these is Newton's second law of motion, which affirms that the force is equal to mass times acceleration of the stone.

Mathematically this law is what mathematicians call a (second-order) differential equation—i.e., an equation that contains second derivatives like $\frac{d^2x}{dt^2}$ —whose solution always requires a knowledge of two constants relating to the present motion. One is the present position, which can be measured directly. The other is its present velocity. But present velocity is determined by noting the present position and somehow determining its position a short time *earlier*, then dividing the change in position by the time interval. We might say that knowing the present velocity requires us to take a small step back into the past. Were we to find the present acceleration (which, as it happens, is not necessary for this task of solving Newton's law), we should have to know the stone's present velocity and its velocity a short time ago. But the latter requires knowledge of the position at a time a little earlier than the one that gave us the present velocity. We would have to take two steps back into the past.

Solving the equation gives us the position and velocity of the stone at any future time: This is a causal prediction because it produces knowledge of the future position and velocity from position and velocity of the present or past. In general a causal prediction involves a law and a specific, limited set of observables, sometimes called causal observables. Thus, so far as causal prediction of motion is concerned, only mass, force, acceleration, velocity, and position are significant. The others also enter into lawful combinations when processes other than motion are considered. In the less developed sciences such as sociology and economics the causal laws are often verbal, sometimes simple algebraic equations involving time.[3] The most important and general causal laws in physics are differential equations.

The point we wish to make is that the principle of causality allows us to predict the future in terms of observables relating to the past. Determination of present states (or events) by future states (or events) is known as action of *purpose*. This statement, however, requires qualification. The causal principles can be reversed. We can use them backward in the following way: If we know the observables of the stone at a future instant for certain, we can use the causal law and retrodict the present and all past states. This illustrates reverse causality but not purpose. The latter comes into play when no strictly causal law is known, and the foregoing retrograde steps must be replaced by steps into the future. Purpose, determination by the future, is not a significant agent in the present physical sciences. It requires a vision of a future goal, and this is possible only for conscious beings. We expect that purpose plays a role in the sciences of life.

We have stated that all theories describing or revealing physical reality are under the guidance of, and therefore satisfy, the principle of causality. Whether or not this is intrinsically necessary and will always be the case has been debated by such philosophers as Kant and Hume. Immanuel Kant took it to be an a priori, a necessary ingredient of human thought, while David Hume regarded it as an incidental, useful tool. As for the authors of this book, we are impressed by the seemingly universal validity of this guiding principle but do not rule out its eventual abandonment, certainly not in the mental realm.

Having illustrated the meaning and the working of causality in connection with a simple case—the motion of molar bodies—let us mention briefly two further instances that show the enormous variety in the nature and complexity of causal observables. The state of an electromagnetic or radiation field has the causal observables: electric and magnetic field strength, permeability, dielectric constant, electric charges, and magnetic poles. Together they comprise a (causal) state. Causality is conveyed by Maxwell's equations (as discussed earlier), which allow the prediction of future states in terms of the above-mentioned observables. In quantum theory a state is a probability distribution with respect to one given observable, usually the position of an *on*. If the position itself, or any of the "classical" observables like velocity, acceleration, or energy were chosen as observables, the causal principle would no longer be true. Our lesson is this: Causal observables are different, often unpredictably different, in different domains of science, even the sciences of physical reality.

Among all the changes that have occurred in the meaning of causality, the one required by the quantum theory is the most radical and the most important (see chapter 8). This can be illustrated with examples, one drawn from the molar realm, the other from the microcosm: The first is the behavior of billiard balls, with which we are already familiar, the second that of electrons.

The causal state of two billiard balls involves knowledge of two sets of observables, the position and the velocity of each. The state is given in terms of twelve single numbers, the x, y, and z-components of position and velocity of the two balls. Knowing their values at the present moment, a solution of the equations of motion provides the values of all twelve numbers at any future time, even if the balls undergo collisions.[4] This example typifies what philosophers call determinism, the view that every phenomenon—physical, ·biological, social, or moral, must have a cause and is therefore predetermined. It is a special form of causality, the most rigorous and the most precise of all its formulations. Until the advent of the quantum theory in the beginning of this century it was thought to be the only type, applicable in principle in all fields of science. To be sure, it was known that there are many scientific and other domains in which strict determinism could not be proved, fields like psychology, sociology—indeed, all areas of human behavior—but the failure of "classical" determinism was ascribed to lack of precision of our knowledge. Laplace claimed that a mathematical demon, a mind capable of solving an extremely numerous and correlated set of differential equations, could predict the future position and velocity of every molecule in the world if it knew these observables at the present time. And he implied that the same would be true for a human body if the demon knew the state of every one of its constituents at present. Strict determinism was regarded as the only valid form of causality; departures from it were ascribed to human ignorance. Indeed, the appeal of determinism was, and is, so great that even today some experts in the theory of quanta and nuclear physics are searching for schemes to restore it by modifying currently accepted theories of elementary onta. We shall return to this point in a moment.

Turning now to a second example, the behavior of an electron, we find it impossible to define its state in terms of its present position and velocity because Heisenberg's principle of uncertainty, one of the revolutionary and incontrovertible historical bases of the atomic theory, forbids the simultaneous measurement, and hence prevents a knowledge, of the electron's present position and velocity. However, another form of causal prediction remains possible. If we use a new definition of "state," this time in terms of a single observable—which we may take to be the electron's position in a wider sense not referring only to a single point of space—we restore causality in a generalized sense, somewhat as follows.[5]

Instead of defining a single position x,y,z in space, we determine a function phi (x,y,z), called the state function, which makes reference to *all* points of space. It does not tell us where the electron is, and it cannot be determined by a single observation. But a multitude of observations does lead to a knowledge of the state function φ.

For the sake of definiteness, assume that a stream of electrons is sent through a succession of slits so that upon emergence from the last slit they all progress in a narrow beam along a known direction. Producing this beam is called the preparation of a state of each of the electrons. To find the state function, our $\varphi(x,y,z)$ above, we must in principle allow an electron of the beam to manifest its presence at every point x,y,z by putting a position-measuring device at every point and recording the presence or absence of an electron. If the beam contains only one electron, we must reprepare the state many times and perform the same kind of measurement after each preparation. In this case we cannot assume that each measurement is performed upon the *same* electron. This is no obstacle to our reasoning, however, for in the microcosm there is no way of specifying the identity of an electron.[6] From these, in principle innumerable observations, the state function $\varphi(x,y,z)$ can be constructed.[7]

It will be clear that φ cannot enlighten us on what to expect in a single measurement. It refers to an aggregate of measurements, a probability aggregate obtained by squaring $\varphi(x,y,z)$. More precisely $\varphi|(x,y,z)|^2$ defines the probability that the electron will be found at x,y,z. This probability says nothing about a single observation; it defines a relative frequency—i.e., the number of times the electron is found at the point x,y,z, divided by the total number of observations.

We now face a new form of causality: If $\varphi(x,y,z)$ is known at the present moment, its altered form at a later time can be computed. In more general terms, a probability distribution known at present allows the prediction of an altered probability distribution at a future time. The causality principle, according to which a present state determines a future state, remains valid if the state function is considered to define a state. Determinism is abandoned, but causality in a probabilistic form remains.

The philosopher wonders why this should be the case. The change is not due to human ignorance; it is a fundamental principle, Heisenberg's, that makes knowledge of a deterministic electron state impossible. This leads us to face the question: Is probabilistic—in other words, statistical—causality more basic, more general than determinism, or is it a degenerate form of it? Based on two considerations, one scientific and technical, the other philosophical, we affirm the first alternative.

As we have already discussed, the probabilistic description of the microcosm congeals to a deterministic theory of the molar world in a purely formal way. Now, of the two types of description the probabilistic one is the more embracive, the more general, for probabilities can "reduce" to certainties, but not the other way around. Certainties are probability distributions, which have the value 1 for a specific value of an observable, the value 0 for all others. This conclusion in turn leads us to examine the history of our knowledge in different domains of science, and more generally in different realms of reality.

But first, a few words on the quantitative meaning of probability. It generally refers to events, such as throwing a die or winning a lottery prize or, in science, obtaining a certain value when performing a measurement. If a die is thrown many times, say 600, the number 2 will appear uppermost about 100 times (for reasons that are irrelevant so far as probability is concerned). On the basis of a long series of throws we then *define* the probability of throwing a 2 as 100 divided by 600—i.e., the number of successes divided by the total number of throws or, more generally, events. If you play 1000 times in a lottery and win, on the average, twenty times, the probability of winning is 20/1000, or 1/50. If you want to know the precise length of an object, you measure it a number of times, obtaining slightly different values each time. Suppose the measured values are 0.89 feet, 0.91 feet, 0.91 feet, 0.89 feet, 0.90 feet, 0.88 feet, 0.92 feet, 0.90 feet, 0.90 feet; they differ because every measurement contains an inevitable error. We then say: The probability that the object has a length of .88 feet is 1/9; a length of .89 feet, 2/9; a length of .90 feet, 1/3, etc., and we take the length having the maximum probability, in this case 0.90 feet, to be the "true" length of the object.

Clearly, if all values turned out to be the same, i.e., if certainty prevailed, that value would have the probability 1, and all others the probability 0.

Until recently all empirical attempts to achieve knowledge were forced to use a probabilistic route, were unable to make absolutely certain predictions. The one exception was "classical" physics, which comprises ordinary mechanics and the theory of electricity and magnetism. In the eyes of nearly all physicists and most other scientists this was *the* most fundamental science to which all others should ultimately "reduce." Despite overwhelming evidence from all life sciences, despite the fact that every physical measurement, and hence every state determination, is afflicted with inevitable errors of observation, the physicists of the last century remained in their ivory towers, proclaiming that they, and apparently only they, were able to make absolutely certain determinations of their observables and could therefore proclaim the doctrine of determinism. Then came Heisenberg and the quantum theory, and even the physicists had to surrender their singular and somewhat boastful view.

Going a little deeper into philosophy we encounter a phenomenon that provides further evidence for the universality, the necessity, not of classical determinism but of statistical causality. It is sometimes called "observer participation" in a measurement, or more properly, we believe, epistemic feedback. Let us explain both of these terms.

The usual explanation of the uncertainty principle of quantum mechanics is this. Two observables, like position and velocity of an *on*, cannot be measured and therefore cannot be known simultaneously because the measurement of one disturbs, alters the value of the other. A well-known example is this. The position of an electron is measured by irradiating it with light (gamma) rays. Their reflection tells where it is. But rays, in being reflected, alter the speed of the electron, so that its speed at the moment of reflection is not exactly measurable. This disturbance is called the observer's participation effect: The observer, while making a measurement, does something he cannot help doing and thereby alters the value to be obtained. If this were the sum of the matter, the word "participation" would be somewhat misleading; for in the first place every measurement, even if it were infinitely precise, would require an observer's participation. The crucial point is that this participation *alters* the result, makes it uncertain.

We think this effect, this unpredictable disturbance, having penetrated into the one domain from which it had been previously excluded, is universal. We believe that *every* observation, no matter what its pretended precision, is subject to an unknown and unknowable disturbance arising from the fact that it is being made. We like to call it "epistemic feedback," a term chosen for the following reason. *Episteme* is the Greek word for knowledge; epistemology means theory of knowledge. Knowledge presumably has to do with being, and the effect under consideration is one in which knowledge, or the act of obtaining it, alters being. Modern science speaks of inevitable interaction between two factors as feedback. Hence we chose the term "epistemic feedback" for the phenomenon under consideration, wherein knowledge alters being. And we believe that its inevitablity merits emphasis in all of modern epistemology.

Examples of this come to mind in great abundance and in the most diverse fields. Knowing that one has a certain disease can make the disease better or worse. Every doctor, every psychiatrist knows that the way a medical condition is determined, the way the patient is asked a question, can or even will alter the condition. Aspects of psychic healing may involve epistemic feedback. Knowledge of guilt alters behavior. If a group of people is aware of an economic or social trend, the trend will often be changed. Epistemic feedback is apparent in predictions, and especially in deterministic predictions based on observation. If an analysis of the present market condition induces the belief that a certain stock will increase in value by a specified amount at a future time, that belief will doubtless be erroneous: The value will outdistance the prediction because more people will buy the stock. If the rate of inflation is "measured" and found to be high, experience shows that the rate will increase if the measurement is believed. Researchers in parapsychology seem to have shown and frequently use the argument that a subject's knowledge of possessing a certain psychic ability usually changes that ability. These are only a few examples.

To summarize the preceding considerations, we recall that causality can have several forms, that different domains of experience, different types of reality display or require different interpretations of cause and effect. Strict determinism is a very limited version of causality. Statistical causality that takes account of epistemic feedback is the most general and most widely applied. We should therefore not be surprised to find that most domains outside of a small area of physics, most types of reality, employ the causal principle in a statistical way.[8]

PART III

Domains of the
Social Sciences

... the checkered, fateful adjustment of man to the outer world. This
ceaseless shifting in man's relations to the impressions crowding in
upon him from the surrounding world forms the starting point for all
psychology on the grand scale, and no historical, cultural or artistic
phenomenon is within reach of our understanding until it has been set
in the perspective of this determining point of view.[1]

IF WE ARE TO USE the new methodology to explore our
world, we must ask basic questions in each domain of experience in a
fresh and unbiased manner. We must not expect our experience in
other domains to indicate necessarily what our experience will be in
a new one. We ask in each domain, "What are the observables here?
What kind of measurements can we make? What laws relating to
observables can we hypothesize and test? What do the terms 'space,'
'time,' 'state,' 'observer' mean in this domain?" The answers may or
may not be very different in each domain. This is what the physicist
Percy Bridgman meant when he wrote:

It is difficult to conceive of anything more scientifically bigoted than to
postulate that all possible experiences conform to the same type as that
with which we are already familiar and therefore to demand that explanation use only elements already familiar in everyday experience.[2]

There *are* certain things we are committed to in advance of our
findings (although we must be aware that we may find realms in
which even these do not apply). These are that the observables in
each domain will be lawfully related to each other. Further, that
there will be lawful relationships between the different domains and
realms. The principles by means of which we establish the existence
of observables are also clear.[3] However, what these observables will
be, or the meaning of "measurement" in each domain, are things we
cannot know until we have made our explorations.

[139]

Max Planck, the physicist who originated the field of quantum mechanics, has used the phrase "a two-tracked universe."[4] This is the same concept, largely forgotten in science, that we have been developing in this book, though it now turns out that the universe is far more than "two-tracked." It is multitracked in exactly the meaning Planck used. There are various "tracks," various realms of experience, each with its own observables, laws relating them, and definitions of time, space, state, and observation needed to make the data meaningful. They are compatible, and each leads inexorably into the others, but they are different, and each must be taken in its own terms.

Descriptions of mental states, social attitudes, significance in art, and like subjects almost always use metaphors—feeling blue, ebullient rage, high ideals, flat painting. Rules of correspondence, operational definitions that were discussed in detail in chapters 3 and 5 are not available for these domains except in a modified probabilistic form. The scientific use of the term "probability" refers, as we've noted, to the relative frequency of a specific observed event—i.e., to the number of times it happens divided by the total number of observations made (ideally an infinitely large number). The social scientist is concerned primarily with molar behavior and man's inner life. In these realms the second law of thermodynamics no longer holds. This is largely because purpose replaces cause in many domains of social science: Too, space and time take on meanings that are often different from those in the see-touch realm. Prediction becomes possible only in terms of probabilities—i.e., in terms of the relative frequencies of observed events.

In the nonphysical realm we can distinguish several types of reality: sensory, clairvoyant, transpsychic, and mythic. In the domains of art, we must pay attention to the intent of the artist, the responses of the beholder, and the domain of man-made things.

Nature, too, presents aspects of beauty. These are not usually regarded as art, but a look at nature reveals a similarity between the idea of beauty as a principle of art in the sense in which truth is a principle of science.

This consideration—indeed, this book—would hardly be complete without a careful assessment of the divergence of views of two geniuses, Newton and Goethe, with respect to the nature of color. Hence we juxtapose and discuss two books: Newton's *Optics* and Goethe's *Farbenlehre* (theory of color). Each of them presents a viable view of the reality of color; Newton is accepted by physicists,

Goethe by many artists. As we shall show, neither view can be said to be wrong, but each serves an important purpose.

Parapsychology, the phenomenon currently designated by psi, is another realm we are now able to deal with. This may be regarded by some readers with "scientific bias" as a daring feat; needless to say, we do not accept all claims that have been made by many authors writing in this field, nor do we regard it as a developed science. However, we believe that there is an important need for research in this area and offer suggestions as to how this might proceed. Our conclusions with regard to the problem of reductionism and our theory of realms and domains are particularly useful here.

Of utmost importance to people of all cultures are the problems of ethics. We may now apply our new viewpoint to the domain of ethics and notice a rarely considered parallelism between the methodology of science and that of ethics.

We will deal finally with consciousness. Here the conclusions reached earlier in this book, especially with respect to reductionism, again become important: Mind processes do not "reduce" to the physics and chemistry of the brain but transcend them in a manner compatible with, and indeed dependent on, physical and chemical actions of the body. This compatibility is exemplified at the end of the book by a resolution of the free-will problem, a solution made possible by recent developments in physics coupled with a recent analysis of the act of volition, made possible largely by Sir John Eccles. Strangely, this result turns out to be a modern version of St. Augustine's proposal that human freedom is a supervention of mental choice upon physical chance, the chance being offered by quantum mechanics.

12 The Domains of the Social Scientist

But it behooves us to be very careful, not to forget that we are dealing only with analogies, and that it is dangerous, not only with men but also with concepts, to drag them out of the region where they originated and have matured.
—SIGMUND FREUD

To RETURN NOW to a theme introduced early in this book, we recall that the social scientist deals with the problem of alternate realities by asking how the individual or the culture under study construes the world at a particular time or in a particular situation. The scientist seeks to determine what is the set of rules that are seen to be operating in the total cosmos—everything "that is"—by his subjects. (In the experience of the social scientist people tend to use one construction of reality at a time. When a person is dreaming, for example, he invests the whole of reality with the laws that govern the dream. When he is trying to cross a busy highway, he invests the whole of reality with an entirely different set of laws. As we have shown in the case of our apocryphal businessman back in chapter 1, this set of observables and laws, seen as operating in all reality, changes at least several times a day for all of us.

The general orientation of social scientists today is that there is only one correct way of perceiving and responding to reality: The "common-sense," Western, everyday way. It is the way our businessman is supposed to be perceiving and reacting while he is at work at his desk. Social scientists generally believe in the single rationality governing the entire universe, and they are usually pretty certain that they know what that rationality is. It is very similar, if not identical, to the rules that the physicist attributes to the see-touch realm: The realm of experience in which all the entities can be at least theoretically seen or touched.

As we look at the history of the social sciences in the past hundred or so years, we see the repeated attempt to apply this model of reality—essentially the machine model—to human behavior and experience. Everything that humans (and animals) did or experienced was to be explained in the same terms as we explain the behavior of machines. The social scientists neglected the words of William James:

We have so many businesses with nature that no one of them yields us an all-embracing clasp. The philosophic attempt to define nature so that no one is left out, so that no one lies outside the door saying, "Where do I come in?" is sure in advance to fail.[1]

Instead of following this advice and trying to understand human behavior and experience in their own terms according to their own laws, the social scientists in general proceeded on the basis that everything they studied could, and must, be ultimately explained in terms of the Western common-sense, machine model of reality. Everything could be explained as if it were governed by the same rationality that governs a machine. This inevitably led to the "nothing but" tendency in the explanations of the social scientists. In this view, wrote Arthur Koestler, human beings are "... *nothing but* [itals ours] a complex biochemical mechanism, powered by a combustion system which engages computers with prodigious storage facilities for retaining coded information." Goethe's great scientific and artistic work was *nothing but* an attempt to cure his premature ejaculation. Your tastes in paintings and furniture are *nothing but* the equivalent of urination on your wall to mark your personal ownership—marking out your territory by covering its boundaries with the smell of your urine. Unfortunately, Koestler was not making up these statements as outrageous examples, but quoting from well-known social scientists of the present period. He went on to say that it would be equally true to say that a human being is nothing but 90 percent water and 10 percent mineral traces. This statement is true, but a greater lie because of what it does not consider.[2]

The implication of the idea of one rationality governing the entire world was the idea that animals and humans can be explained in the same terms and dealt with in the same way. This implication was responsible for the anthropomorphizing of animals—viewing them as if they have the same experiences and behave in the same way for the same reasons as do human beings. Thus, in the Medieval period, there were frequent trials of animals that had committed crimes. A full legal action would be brought against an animal, and it would be condemned to the same punishment that a human being who committed the same act would suffer. One branch of the social sciences still operates on this theory—it tries to explain human behavior and experience on the basis of animal experimentation. Since humans and animals are both essentially the same, and since it is often difficult and problematical to experiment on humans, it makes sense to work with animals. Yet, after more than half a century of intensive experimentation on white rats and pigeons, we find it doubtful that anything of value about human behavior or feelings has come out of the animal psychology laboratories.

With the growth of the machine concept and the increasing success of science in understanding how machines worked—making their behavior lawful—a new problem arose. Human beings and animals obviously had characteristics that machines did not. It was not enough to go from anthropomorphizing animals to bestializing human beings; neither was governed by the machine model, despite the basic assumption of our culture that everything can be and must be explained in the same terms. Gradually early in this century the paradox was solved. We would *mechanize* living creatures. Human beings, rats, machines, pigeons, graylag geese, they all have the same characteristics and behave in the same manner for the same reason. This led to what L. V. Bertalanffy (the founder of the modern science of Systems Theory) has called:

... the robot model of man ... It ... led to a remarkable degree of theoretical unification. Machines, animals, infants, and mentally sick provide adequate models for human behavior. Machines, because behavior is eventually to be explained in terms of machine-like structures of the nervous system; animals because of the identity of principles in animals and human behavior and because they can be better handled; and infants because in these—as well as in pathological cases—the primary factors are better recognizable than in the normal adults.[3]

Despairingly, D. H. Lawrence wrote:

"Oh, I have loved my fellow men,
And lived to learn that they are neither fellow nor men
But machine robots."

But in developing the model of man as a machine, the Behaviorists in the first half of this century ran into the problem of the human activities of thought and language. Since machines did not think, it must be that humans did not either. However, machines did make noises that sometimes communicated their internal processes. A power loom clicking along steadily is signaling quite a different message from the same loom loudly and erratically squeaking and clanging. Therefore, one *could* conceive of human robots talking. When people "thought" that they were thinking, then (since this was impossible in the model decided on in advance), they were really just talking so quietly that no one else could hear them. Thought thus became subvocal speech, tiny movements of the voice box. At one time there was a debate over this matter between the historian Will Durant and John B. Watson, the founder of Behaviorism. In the middle of the debate Durant put down his notes, turned to the audience, and said: "There is no point in going on with this discussion. It is apparent that this is a matter about which Dr. Watson has already made up his larynx!"

As we have repeatedly indicated, the essence of the organization of reality we are presenting here is that data in each realm of experience must be taken on their own terms without preconception. It is not only the observables and their relationships that may differ in each realm of experience, not only the definitions of space, time, state, and observer, but the very *methods of study* appropriate to each domain may differ. All methods are not suitable to all realms of experience, but each may be the only appropriate method for one or more.

Once it has been admitted that human behavior has its mechanical aspects, then it ought to be obvious that these are the aspects that the methods appropriate to the study of mechanism will reveal. If you study man by the method suited to chemistry, or even if you study him in the light of what you have learned about rats and dogs, it is certainly to be expected that what you discover will be what chemistry and animal behavior have to teach. But it is also not surprising or even significant if by such methods you fail to discover anything else.[4]

What the mechanists and behaviorists have not understood is that it is perfectly legitimate, and indeed even necessary, to be anthropomorphic about human beings. "How else is Anthropos to be studied, except in anthropomorphic terms?"[5] In studying man the mechanist prefers mechanomorphism or rat- (or pigeon-) omorphism to anthropomorphism. He denies human beings as human beings and says instead that they are machines or rats and should be studied that way. He then reports that in studies appropriate to rats he has found nothing that is not ratlike, and therefore nothing nonratlike exists in human beings. It is as if a carpenter threw out all his tools except a hammer and then insisted that everything must be treated as a nail. If we wish to make progress in the social sciences, it will be necessary

... to dismiss the mechanist's *a priori* assumption that only a certain kind of evidence, collected by certain arbitrarily restricted methods, is really valid. There is an Idol of the Laboratory as well as of the Market Place. And we can emerge from the error which it fosters only if we cease to believe that a thing is obviously an illusion unless it can be measured and experimented with by the same methods which have proved useful in dealing with mechanical phenomena.[6]

The mechanistic psychologist denies free will, partly on the ground that you cannot have a science if things "hop about" to use a phrase of the protagonist of Skinner's *Walden 2*. He cannot accept the idea that science can be based on anything but specific-event predictability, with everything being fully determined. Without the Laplacian view, he feels science will fail. The assumption—which collapsed with the new physics—that if you have enough data, you can know exactly what is going to happen, was a comfortable one.

One of the great attempts of the social sciences to apply the concept of a single rationality was the work of Karl Marx. In effect Marx viewed society as a great machine that would follow a rigidly determined course, although its speed of operation was somewhat flexible. The same factors that applied to the "medium range," the see-touch realm, were applied to society as a whole. The idea seemed reasonable and made excellent "common sense" to very large numbers of people in spite of the fact that many of the major predictions arising from the idea have proven false (communism occurring first in agricultural Russia rather than in the predicted industrialized Germany, no mention of fascism in *Das Kapital,* etc.). Marx's work is still seen by many as a true blueprint of the great engine of society with its inevitable course of development built into its specifications. It is "truth" because there is believed to be only one truth, and since we know this truth in the machine, we can also perceive it in all other realms, including that of social development.

The machine model of man was already being described in the 17th century and reached its full flower at the end of the 19th and beginning of the 20th centuries. In 1662 Descartes published *De Homine,* a theory of man and animals as machines. He even made a theoretical model of a mechanical man. In his view all animals were completely mechanical, although man retained an immortal soul. In 1747 La Mettrie went the whole way and made man completely mechanical in his *L'Homme Machine.*

By the beginning of the 20th century every major model for the explanation of human behavior was mechanical. Darwin used the machine model for evolution, although he added the refinement of the machine flipping pennies in his "accidental variations" in structure. (This is an entirely legitimate and consistent concept in the machine model—there is no reason why a machine should not throw dice or flip pennies in determining its next action.)

The ubiquity of the model is shown by the fact that Darwin's work and his concept of Natural Selection were not only seen as a justification by the communists, but also by the proponents of laissez-faire capitalism and of fascism. (Marx believed that he owed a debt to Darwin and wanted to dedicate *Das Kapital* to him. However, permission was refused.) So great has been the belief that the machine model was the correct and inevitable model that its proponents have always seen more validity in their predictions than the facts would indicate was there. They have been "dizzy with success," as Stalin said in his famous speech celebrating the triumphs of rural collectivization on the eve of the great famine of 1932, in which over 4,000,-000 Russians died of starvation.[7] Psychoanalysts, Behaviorists, and Darwinian evolutionists alike have also generalized their ability to explain and predict far beyond the degree the data would indicate as valid.

Psychoanalysis was seen as a complete system that could explain and deal with all aspects of human behavior. What was ignored except as temporary lacunae in the field of knowledge—small gaps to be filled in later—was the fact that it was inadequate to account for creativity or for the long, slow struggle *up* from the caves that the human race has made. Psychoanalysis could not account for the beauty seen in a sunset, for the genius of Mozart, for the opera that Freud loved, or for his courage and his devotion to humanity. To apply psychoanalytic methods and interpretations to Freud himself would be an insult to that profound and suffering giant. For the psychoanalyst practically everything that the patient said was a symbolic reference to something else; the idea that the patient was saying what he actually meant (at least in the first several years of analysis) was seen as a very naive point of view. Sexual symbols were everywhere. During a panel discussion of symbolism Will Durant was once asked what he thought of the Freudian interpretation. He replied, "It occurs to me that just about everything in the world is either sort of long and pointed, or sort of round and hollow!" (And even Freud once remarked, "Sometimes a cigar is just a cigar.")

When we note that these major social explanation systems of our time all used the machine model, we can see how firmly this idea—

that the whole universe was run on one system—underlies the thinking of this period. One might say that perhaps the greatest discovery of the 17th century was that the model of reality produced by examination of the see-touch range could be used everywhere and explain everything from atoms to societies. The greatest discovery of the 20th century seems to be that it can't. Norbert Wiener has stated this emerging discovery of our century as a new commandment: "Render unto computers the things that belong to computers, and unto man the things that belong to man."

Again we must point out that this idea, that all the world can be described in the same way, arose only after the concept of one God had become firmly established in the West. The Greeks saw matters quite otherwise.

The positive insight of Socrates was that we cannot discover the nature of man in the same way that we can detect the nature of physical things. Physical things may be described in terms of their objective properties, but man may be described and defined only in terms of his consciousness.[8]

In a statement whose importance we are just rediscovering in the present century, Aristotle warned us:

It is the mark of a properly trained mind to look for a degree of precision that is appropriate to the subject matter, and only to the degree that the nature of each allows.[9]

The strength and pervasiveness of this assumption—that the entire world, including human activities, can be described only by the mechanical model—in our thinking is difficult to overestimate. Neither hard facts (as the examples mentioned above of wrong predictions in the bible of communist theory) nor the inability to account for clear data (as psychoanalytic theory cannot account for creativity, love, courage, or dignity) could disturb its secure position in our thinking. It seemed to be such obvious common sense. If everything works on the same principles, and we know the principles on which one thing works, we know the principles on which everything works. We commit ourselves to them in advance of our doing research. And the model was the sensory realm in which science had made such spectacular advances.

... physics has advanced so rapidly that it has come to be regarded as the standard and basic science. We tend, therefore, to assess degrees of "reality" in our thinking, according to its conformity or not with the laws derived from our study of physics. "Make me a model of it and I will be prepared to accept it" is not only the demand of the Kelvins of this world, but of ordinary men and women.[10]

But in domain after domain of the social sciences, the model failed to produce anything like the predictability and control that had been achieved in the realm from whose observation it arose. Even in animal psychology, with the exception of rats and pigeons, we could find nothing like precise mathematical equations connecting stimuli and meaningful units of behavior. The animals did not seem to follow the laws we insisted were there. Indeed, among animal psychologists there began to spread a scandalous and heretical rumor that the First Law of Animal Psychology is really as follows:

If an animal of known, stable genetic background is raised in a carefully controlled laboratory environment, and administered a precisely measured stimulus, the animal will act as it damn well pleases.

The theory that man was a complex machine whose behavior could be predicted and controlled, which was largely accepted by the social scientists, became curiously nonfalsifiable. Each failure at prediction and control was taken as evidence that we did not yet know enough to implement a correct theory. Each success (as of an I.Q. test or opinion poll) was taken as proof that the theory was correct. The social sciences became largely pseudo-sciences as their basic hypotheses could not be disproven.

The day before the 1980 presidential primary in New York State, the Harris Poll (the most expert and respected of the modern prediction organizations) reported that its studies showed a strong victory for then-President Jimmy Carter. The next day Senator Edward Kennedy emerged the victor by a sizeable majority. No one suggested that there might be a basic flaw in the theory of behavior prediction techniques; it was assumed that the error was due to a corrigible flaw in prediction technique. Indeed, it is hard to conceive of any experiment or any failure in prediction that would cause really basic questions to be asked about this issue. This is the definition of a pseudo-science; a nonfalsifiable field of inquiry. It is very difficult to conceive of an occurrence or an experiment that would convince a psychoanalyst, an astrologer, a Marxist, a Behaviorist, or a Darwinian that the concepts of their school were invalid.

The social scientist who believes that a human being operates on the same principles as a machine is in a very peculiar position indeed. For if his behavior is absolutely determined by his past, if he has no free will and no goal-oriented behavior, no "purpose" (and these things *cannot* exist in the machine model), then how seriously should we take his pronouncements? His conclusions are not freely determined by the data he is studying as he insists they are, but by his conditioning (early or late in his development, depending on the deterministic school you follow) and so why should we take them as of any value? When the social scientist of this persuasion says that he has made an objective examination of the data of his field, and come to the conclusion that there is no such thing as free will or objectivity, and that all perceptions and conclusions are determined by conditioning, we must apply these same standards to the social scientist himself and dismiss his work. (He is very much like the psychoanalyst we find in front of his extensive library who assures us that one cannot change one's behavior by reading, and that gaining new information from books or other intellectual endeavors will not really affect our experience or behavior. Similar also is the philosopher in existential despair who writes book after book to convince us that human beings cannot communicate with each other.) One writer summed up this paradox:

If thought be merely cerebral chemistry, and the theory of Behaviorism a conditioned reflex, neither materialism nor Behaviorism need to be considered seriously. Like the scorpion who is reputed to sting himself to death when surrounded by a ring of fire, both theories commit *hara-kiri.*[11]

To paraphrase Schopenhauer, it seems legitimate to say that Behaviorism does not need a refutation but a cure. The philosopher C. D. Broad once referred to it as "one of those systems so innately silly that it could only have been devised by very learned men."

Much of academic philosophy follows the implications of the theory of one rationality governing the cosmos. The Logical Positivists set themselves the task of building a language that would provide a foundation for all the sciences and reflect "truth" and "reality." They believed that there was a single language and a

. . . unique model for all real science and that—when they had described it—they would verify all science. Ultimately it would verify all experience.[12]

After over half a century of effort it has now become clear that no one language can do this. Not only are different metaphysical systems necessary to describe different realms of experience, but often different kinds of language are needed to describe experiences in these realms. These necessary languages vary as much as computer language varies from Beethoven. Being in love demands a different kind of language to describe the experience than does the Kinsey approach to making love. They simply cannot be both adequately described in the same system of communication. At one time the philosopher Gabriel Marcel was lecturing to a group of American Logical Positivists on grace and transcendence. They kept telling him to speak more clearly and to "say what he meant." Finally Marcel paused and then said, "I guess I can't explain it to you. But if I had a piano here, I could play it."[13]

The realms the social scientist studies have quite different Basic Limiting Principles and follow very different laws than does the see-touch realm from which we draw our ideas as to what constitutes common sense. To understand this, we look at the data of the social scientist and the realms into which they fall. We shall follow the method of the physicist in our study of how the social sciences relate to the problem of alternate realities. We shall look at the realms of experience from which the data are drawn and ask, "What kind of measurements can we make in these realms? What are the observables here? What kind of laws can we postulate relating these observables to each other?"

The data of interest to the social scientist fall into two general classes. The first of these concerns meaningful units of behavior—"molar" behavior as opposed to "molecular" behavior.

A molar behavior is: the student's attendance at class, the lecturer's delivery, the pilot's navigation . . . Mr. Babbitt's flirtation, Galileo's work which revolutionized science, the hunting of the hound and the running of the hare, the biting of the fish and the stalking of the tiger. In short, all those countless occurrences in our everyday world which the layman calls behavior.

Molecular behavior, on the other hand, is something quite different: the process which starts with excitation on the sensory surface on an animal, is conducted by nerve fibers . . . and ends in a muscle contraction or gland secretion.[14]

The author of these paragraphs warns us that it is very easy from this to go on and say that since molar behavior always implies molecular behavior (muscle contractions and the like), only these are real, and a true science of behavior will have to study only nerve excitations and muscle contractions, regarding a molar behavior as a sort of secondary and almost accidental process. To say, in short, that one domain of experience is more "equal" than another. This is the error of "reductionism."[15]

The first class into which the data of the social scientist fall is molar behavior. The second is our inner experience. How do we actually experience ourselves and the world? The experience is literally our consciousness of what we feel is going on in our thoughts and feelings. These are real data, composed of observables we clearly and unmistakably experience. We will ask the same questions in this realm as in the realm of molar behavior. Here too we will find that the Basic Limiting Principles, the observables and the laws relating them, are very different from those which we find in the see-touch or "sensory" realm.

It is obvious but perhaps worth emphasizing that just as it is true of observables in the microcosm, it is true that consciousness has no visual characteristics. It cannot be seen or touched, has no color, shape, size, texture, or locatability. We should therefore expect that identity and interaction modes might well be different from those in the see-touch realm, that causality might have a different meaning, and that it might not be possible to make a mechanical model of consciousness or parts of it. The criteria for reality developed for the sensory realm are valid for it but are not necessarily valid outside of those domains in which things can be seen or touched.

Before we examine the specific differences between the observables, measurements, and laws in the realm of molar behavior and of inner experience from those in the see-touch realm, there is one general difference we must mention first. This difference is so crucial that, in itself, it forever removes any hope of explaining molar behavior or inner experience in terms of the same principles as those of a machine. It shows clearly that we are in different realms with different organizations of reality necessary to make the data lawful.

There is a general law covering events in the see-touch realm. This is that unless something special is done from outside of whatever system you are using, things inside the system tend to get less and less organized and specialized and more and more diffuse and everywhere the same. If you heat the tip of a knife in the flame and then turn the flame off, you have a knife with a very hot tip, a warm blade, and a cool handle. Things are highly organized, so to speak; they are different in different places. Once the flame is off, however (and nothing more is being done to the "system" you are considering—the knife), this general law begins to operate. The tip of the knife gets less hot, and the blade gets hotter. Presently the whole knife is the same temperature—the heat has become more "diffused." The process continues: The air around the knife gets warmer, the knife cooler. Presently the air in the room and the knife are exactly the same temperature. The heat has become even more diffused. This theoretically continues until the air in the room and the room walls and ultimately the outside environment are all the same.

This law—that unless something is done about it, things get less and less organized—is a law of profound significance in the middle range. It covers *everything* in that range. It takes extra energy to keep things from getting less organized and less articulated, and this energy must come from outside of whatever system you are using. This is one reason that any patent office in the world will automatically reject your application if you try to get a patent on a perpetual motion machine. There is no such thing in the see-touch range because you always need extra energy from outside your system (your machine) to keep it running.

In physics this is called the Second Law of Thermodynamics. It states that diffuseness, "entropy," constantly increases. Every place gets more and more diffuse. The amount of "entropy" in a system is a measure of how unorganized and diffuse the system is, and it constantly grows greater. This is a rock-bottom, unshakeable law of the see-touch realm. No scientist would question the validity of this law and its universality in this realm.

However, in the realms the social scientist studies—the realms of molar behavior and the realm of the inner life—this law *does not operate*. If we look at the results of human molar activity, we see that the opposite is true. The amount of gold in the world, which in the beginning was scattered through the crust of the earth, gets more and more concentrated in purer form in specific spots (like Fort Knox) throughout the world. The gold becomes more and more articulated in its purity and its location. It becomes *less* diffuse. The process has *negative* entropy. The reverse of what the Second Law would predict has been occurring over the centuries.

"Heat" and temperature are other examples. Think of the millions of stoves and furnaces and refrigerators that exist for the specific purpose of concentrating "heat" or "cold" in specific spots. (And small concentrations of ice appear in glass containers when the weather is hot and sultry!) This is the direct opposite of what the Second Law would predict. Human molar activity is an anti-entropic activity, and in this realm the Second Law of Thermodynamics does not operate.

In the realm of inner experience this is also true. "Order" in this realm, in our inner experience, is in terms of information. If our information is highly organized, the bits of it well related to each other in coherent patterns, we say it is articulated, specific, and the entropy is low. If the bits of information we have are not related to each other in patterns by other bits of information, we say it is diffuse and the entropy is high. Thus the more *patterned* our information, the lower is its diffuseness and its entropy. As a matter of fact, the mathematical formula for information in the science of information theory is exactly the same as the formula for negative entropy in physics.[16]

In this way, when we speak of entropy in the realm of our inner experience, we are speaking of the organization or disorganization of our information—our knowledge. As we look at this, a strange fact emerges. If I have three facts about something and you give me another fact, this does not necessarily make four facts. I may put the new fact with the older ones I have and find that I now have five or seven or twelve facts. And this new information, this new order, *was not achieved at the expense of the outside environment.* So far as the Second Law goes, I can only achieve increased order by putting something new into the system from the outside. (Thus when I wanted to increase the specific differences of parts of my knife in the earlier example, I had to turn on a flame from outside the knife.) In the realm of inner experience, however, the law is frequently violated. I may even put together two bits of information I have had for a long time and find that this gives me a large number of highly organized new bits. I have understood something new about the bits I have and see them in a new and more highly organized and articulated way. This is what happens in the creative act.

(Some enthusiastic computer experts are likely to object to these remarks by claiming that a computer could be built that automatically enlarges the number of bits as does the mind. N. Wiener, in a private conversation with one of us, smilingly denied this possibility. But even if it were a fact, we could still not prove that the computer had consciousness and "knew" what it was doing.)

The repeated violations of the Second Law of Thermodynamics in the realms of the social scientist's interest makes it plain that these realms operate according to very different laws than does the see-touch realm. Thus they require a very different organization of reality to make the data in them lawful. It is, from this difference alone, clearly impossible to use the laws of the see-touch realm, which we tend to regard as "common sense," to explain and deal with the phenomena of molar behavior and inner experience. Like it or not, the social scientist is sooner or later going to have to come to terms with this fact and, accepting it, give up once and for all his repeated attempts to apply the common sense, machine model to molar behavior and to human experience. He will ultimately have to listen to Thomas Carlyle's response when he heard Margaret Fuller make her bravura statement: "I accept the universe." Carlyle said, "Madame, you'd better."

Another crucial difference between the realms in which the social scientist works and the see-touch (sensory) realm is the presence of a particular observable—purpose. In the latter realm the state of a system at the present time determines what the state will be at a later time. In other words, a machine does a particular thing because of its structure—its parts and the relationships between them—now: What will happen in the future, what will be the end result of its action, is completely irrelevant. "Purpose" does not exist in this realm. It is not a factor determining what will happen. The piston moves because a spark has exploded the gasoline in the cylinder, not because it also wants to get to a higher position on the cam shaft. "Causality" in the sensory realm is the present state of things, and that is all there is to it. The end result of the action has no effect on the action, which occurs only because of how things are at present.

In the realm of molar behavior the end result of an action partly determines what occurs. The student walks down the street in order to get to the class. Ignoring the fact that his walking has a goal—and that this *purpose* is one of the factors in determining whether or not he walks—would be ignoring an important observable. It would also be plain silly. If I lift my fork with a piece of egg on it, part of the reason for the action is that I want to put the egg in my mouth. (If a derrick lifts a scoop of earth, it does so because of the interaction between the motor, the drum, the cable, and the scoop. It is completely irrelevant to the derrick whether the earth is being put into a truck or dumped into the bay; whether the lifted earth will help to build a school, a fort, or a prison. The *goal* plays no part in the derrick's action, but it plays a major part in the actions of the derrick operator and in the actions of the man eating breakfast.)[17]

It is perfectly reasonable to say that there are no teleological factors playing a part in the operation of a machine or anything in the see-touch realm that is not alive (teleological meaning "goal oriented," that future states influence present occurrences). It is perfectly unreasonable to say that they do *not* play a part in molar behavior. To do so runs against all our experience and ignores a major part of the data in the realm of molar behavior and of inner experience. This arbitrary type of selection is not permissible in science. One is not permitted to ignore observables at will.

Thus in the area of *causation* there is a major difference between the spheres of interest in the social scientist and the see-touch realm. A difference also exists in the definitions, of time and of space, that make sense—enable us to make the data lawful—in the different realms.

The fact that "purpose" exists as a causal factor in the molar behavior and inner experience realms, and does not in the see-touch realm, immediately makes for a major difference in the nature of time in these realms. Time is not "Newtonian" time, which flows everywhere at a steady rate in one direction inexorably. Molar behavior is determined in part by how the individual perceives the future. (For example, as Mark Twain put it, the knowledge that he will be hung in three days concentrates a man's mind wonderfully!) This is already a tremendous difference from the possibilities of the nature of time in the see-touch realm. In addition, time does not flow evenly in our inner experience or in the determinants of our molar behavior. I may feel closer to, and more involved with, the death of my father, which happened many years ago, than I do to the death of President Truman, which happened much more recently. My inner experience and my molar behavior are much more affected by the first than they are by the second. There are qualitative differences between clock time and the time of inner experience, between time measured and time lived.

One of us has elsewhere shown that the perception of time is even different in various social classes in the United States.[18] We can summarize that lengthy analysis, which demonstrated that cultural variations within a society could produce marked differences in how time was perceived and reacted to. Using the class definitions of the "Chicago School" of sociologists,[19] it was possible to show that the "lower-lower" classes built most of their meaningful behavior sequences on the idea that only the present was real and the future had very little meaning (major admonishment to a child—"Stop that

right now or I'll hit you"). The "middle" classes built meaningful behavior sequences on the basis that present behavior was determined
by future events (major type of admonishment to a child—"Stop that
or you will never get into college, or get married, get a good job,
etc."). The "upper-upper" classes built most of their meaningful behavior sequences on the basis that the past plays an important role in
determining the present and future (major type of admonishment to
a child—"Stop that, your grandfather [or "original ancestor"]
wouldn't like it"). The first group ate when they were hungry, the
second at clock-determined hours, the third at traditional hours.
Thus even within a single culture major differences in the perception
of, and reaction to, time exist. Time in inner experience and in the
structure of meaningful behavior is personal time, not "objective" or
Newtonian time. Personal time, however, in those situations in
which we need consistency or the ability to define time periods with
other people, includes clock time. Clock time is thus a special case of
personal time.

We have earlier pointed out the difference between Euclidean
space (the space of the see-touch realm) and personal space (the
space of the realms of molar behavior and of inner experience). The
psychologist Kurt Koffka[20] has demonstrated in some detail how important this differentiation is in the understanding of molar behavior
and human experience. In his terms he distinguished a "geographical" and a "behavioral" environment. The hunting hound and the
fleeing hare are in the same geographical field but in two entirely different behavioral fields. Two siblings live in the same geographical
house, but in the behavioral house of one there is an older brother, in
the behavioral house of the other there is a younger sister. Two men
at a diplomatic party are talking. They are two feet apart. One, an
Englishman, feels that his personal space is being invaded, and he
keeps backing up. The other, an Italian, feels that they are too far
apart really to communicate and remain in contact, that his personal
space is not in contact with the Englishman's personal space, and
keeps coming closer. One advancing, one retreating, they do a
strange ballet down the length of the room, turn around, and cross
the room again. They are in the same geographical space but entirely
different behavioral spaces. These spaces bear no relationship to the
Euclidean space that is the only valid space in the see-touch realm
and that is used by machines.
The geographical environment changes or remains stable accord-

ing to changes in the see-touch realm. The geographical field in which the hound hunts the hare may become wetter or drier with the weather, overgrown with trees, crossed by a highway, and so forth. The behavioral environment changes not only with our perception of such changes as these, but also as our consciousness itself changes. The philosopher Ernst Cassirer once took his ten-year-old daughter to her first opera. As they left *Figaro*, the little girl was deeply surprised to find that her beloved Berlin was very different than it had been when they had entered the Opera House. Everything had changed. The profound difference between the two cities affected her so strongly that she remembered it for the rest of her life.

If I wish to understand the data from the realm of the social scientist, to make the data lawful, I will have to give up the idea that I am dealing with the same definition of space that I use when predicting the movement of billiard balls. Personal space is meaningless to these movements, but it is very meaningful to molar behavior and to inner experience. If I am sitting on the beams of the second story of an unfinished skyscraper, that forty feet in the air is high for me. For a high-iron worker, it is very low. If you and I are the same ten feet of geographical distance from the ocean and you fear the surf and I love it, then if the waves are high, we are in very different personal distances from the water. Machines do not have personal spaces, and their "behavioral" space (if the term has any meaning in this context) is identical with their geographical space. Two bulldozers ten feet from the ocean are the same distance from the ocean, and any attempt to make the data from them lawful will have to bear this in mind. Two people ten feet from the ocean may or may not be the same distance from the water so far as their inner experience and molar behavior go, and any attempt to make the data from them lawful will have to bear that in mind. As it became clear in the 1800s that Euclidean space is a special class of geometric space, it has become clear today that geometric space is a special class of human space. As a human being I perceive and react to *human space* which includes—for certain conditions such as those situations in which I need consistency—geometric space. Geometric space is thus a special case of personal space.

As we have indicated before, the type of predictions that can be made in a realm may differ from the kind that can be made in the sensory realm. In the realms of molar behavior, as well as in the realm of inner experience, prediction is probabilistic and comparative, never absolute. We can never predict the occurrence of the specific event or inner experience, but we can predict that the chance of its occurring is greater in some situations than in others, or that with some individuals the likelihood of their occurrence is greater than with others. Thus we cannot say certainly whether or not a man will commit suicide, but we can say, "Because he is a Catholic, he is less likely to commit suicide than if he were a Protestant." We can say: "Because he is a business executive, he is very likely to be wearing a tie at weekday noontimes." Or: "Because she is an artist and responds to visual stimuli, she is very likely to notice the first changes in the autumn foliage." Absolute prediction of acts and experience of this sort is impossible since these realms do not permit it. After the event the act or experience can be shown to have been determined and inevitable. Before the event it cannot be predicted. This is also true for creative syntheses and ESP occurrences.

One of the questions that arises early in the study of each realm is, "What kind of language can be used to describe the data in this realm?" In the see-touch realm everyday verbal language is generally useful and adequate, although there are situations when some mathematics is needed to round this out. In other realms, however, this is not necessarily so. For example, in the microcosm everyday verbal language is not adequate to describe the data. When one attempts to use it, then the data and concepts in this realm are severely distorted. For example, if we said, "The electron in the hydrogen atom moves in a circular orbit," the implication would be inaccurate.

We have developed specialized languages for the microcosm and for the too large and fast realm. Everyday language is usually adequate in the see-touch realm and probably in the realm of molar behavior. In the realm of inner experience we have never developed a language relevant to the data. We constantly use metaphors from the see-touch realm as if the data of our inner experience were the same as that of our eyes and touch organs. The reason is, of course, that there are no rules of correspondence (see chapters 3 and 4), no opportunities for measurements, for quantification of the ordinary kind in this domain. A partial list of these metaphors might be as follows:

High spirits, low or blue feelings, solid character, stiff attitude, bright outlook, dark gloom, glowing pride, ebullient joy, high ideals, low character, high ambition, deep thoughts, firm or frozen conviction, open or closed mind, calm disposition, boiling rage, burning desire, sharp reasoning, dull mind, soft heart or emotion, shaking or quiet faith, black mood, iron will, bright hope or expectation, red anger, green envy, sparkling humor.

These metaphors are clearly helpful in our attempts to describe and communicate our inner life. However, they leave a great deal to be desired. If the philosopher Condillac was right when he wrote, "A science is a well made language," one of our tasks in developing a science of inner life will be the development of an adequate language.

It is interesting to note that in the exploration of the inner life and of molar behavior the richest language that we have was not developed by science, but by a pseudo-science—astrology. The descriptions of personality, feelings, and behavior used in this technique are far more extensive and intensive than are those from psychology. This appears to be largely due to the fact that astrology took its data on its own terms—from the viewpoint of the data rather than from the preconception that the observables and the laws relating them would fit the conceptual scheme of the see-touch realm. However, while the *vocabulary* developed by astrology is useful, astrology itself is a pseudo-science with no other particular value (except, perhaps, as a way of studying the acceptance of nuttiness in a widely divergent population). This is because it consistently uses both the mythic mode-of-being and the sensory mode. Further, it is—as is general in fields that do this—nonfalsifiable.

The following chart may be of help in showing some of the similarities and differences between the realms of experience which we have been discussing.

We have in this chart barely scratched the surface. First there are almost certainly other realms of experience than the five we have presented here and these others would have in them entirely different observables, laws relating them, and Basic Limiting Principles. Also among the five realms we have presented there are many differences we have not discussed. What, for example, of the *need for meaning*, which we find so strong an observable in the realm of our inner experience and which does not exist as an observable in the realm of interest to the physicist?

Nor have we discussed the fact that the principle of contradiction (A is either A or Not A; a thing either is or it isn't), which operates in the see-touch realm, does not operate in the realm of inner experience. I can, and frequently do, weep as Mimi dies, even though I knew perfectly well that *La Boheme* is fiction. People send all sorts of letters and gifts to characters in soap operas, even though they are well aware that these are fictional parts played by actors. (They do not send gifts to the imaginary town in which the character is sup-

Other Principles in Some Different Realms of Experience

Realm / Operation	Very Small	See Touch	Very Big	Molar Behavior	Inner Experience
Second Law of Thermodynamics applicable	Yes	Yes	Doubtful	No	No
Presence of purpose as an observable	No	No	No	Yes	Yes
Space	?	Euclidean	Cannot be considered separately. Merge into "Space-Time"	Personal	Personal
Time	Newtonian	Newtonian	" "	Personal	Personal
Everyday language can be used to communicate	No	Yes	Yes	Yes	No
Possibility of Replicating Experiments	Statistically only	Yes	Experiments are mostly "thought" experiments validated by observations	Statistically only	No

posed to live; they send them to the television station.) Further, we have not discussed the domain of "life." In this domain there are major factors that make it impossible to use the machine model. For example, machines can be broken up and reassembled and function as they did before. This is not true of things that are alive. Machines produce entropy, and life produces negative entropy. Further, the observable "purpose" appears in life classes. In the words of the biologist E. W. Sinnott:

A remarkable fact about organic regulation, both developmental and physiological is that, if the organism is prevented from reaching its norm or "goal," in the ordinary way, it is resourceful and will attain this by a different method. The end, rather than the means, seems to be the important thing.[21]

One rather interesting conclusion frequently—if illogically—reached from (or at least seemingly legitimized by) the concept of one reality governing the entire cosmos is the idea that "human nature" is the same everywhere and everywhen. This generally seems to lead naturally to the belief that human feelings and behavior cannot really change, although human techniques can. This frequently seems to lead to extremely pessimistic conclusions. In the light of modern anthropology and sociology there seems to be very little evidence for this basic concept of a "given" and inexorable "human nature," and a very great deal of evidence against it. See, for example, Ruth Benedict's classic *Patterns of Culture*.[22]

We have been discussing the problem of alternate realities, as it affects the social scientist, by analyzing the domains and realms from which his data come. This is the way that the physicist approaches the problem. Social scientists, however, use a different strategy. As we have indicated earlier, they analyze the way their subjects are organizing their total experience at a given moment. Although there are innumerable variations, these ways of constructing the world—organizing one's perceptions and reactions—tend to fall into four general classes. In chapter 1 we described an example of each of these as they were used during a day of our imaginary businessman: when he was working at his desk (the sensory reality); when he was praying (the transpsychic reality); when he was dancing (the clairvoyant reality); and when he was dreaming (the mythic reality).[23] We now look in general terms at this approach.

The Roman mystic Plotinus wrote that a human being is like an amphibian that needs to live in both water and on land to achieve the fullest potential. If it lives in only one of these, its development is stunted. (For Plotinus this analogy referred to the clairvoyant and the sensory realities.) In a curiously similar vein the Indian mystic Ramakrishna wrote that a human being was like a frog. In youth as a tadpole, he could grow well in one medium. Later, "when the tail of ignorance dropped off," he needed both land and water for his development.

As we have stated, in the past it has seemed clear and obvious to most mystics and most scientists that one approach (their own) was "correct" and "right," and the other was an incorrect model of reality being used for unconstructive or trivial ends. From the viewpoint of Domain Theory the "either-or" quality of the problem disappears. Mysticism (the development of the ability to use the clairvoyant and transpsychic realities) is one way of modeling and construing the universe. It is suited for some problems and not for some others. Science is another way of modeling and construing the world. It is also adapted to some problems and not to others. Which is correct? It depends on the problem you wish to solve.

Some years ago the child psychiatrist Annina Brandt lectured to a group of rather orthodox psychoanalytic therapists. As she went on talking in a warm and loving way about the experience of being a child, and what the world was like to a disturbed child, the group became more and more upset. Finally one of them asked her, "Dr. Brandt, what *school* do you belong to?" Brandt paused, obviously a bit confused. Finally she answered, "But how can I know until I see the child?"

When seriously worked with, the two (mysticism and science) have much in common. It was not through lack of knowledge that the philosopher Josiah Royce once said that mystics were ". . . the most thoroughgoing empiricists in the history of philosophy." Nor is it an accident that both mysticism and science have arrived at the idea that different constructions of the world are needed for different realms of experience and that these different metaphysical systems must be compatible with each other. As modern science has worked hard to understand the relationships between the various realms with which it deals (e.g., the microcosm, the sensory realm, the macrocosm), so mysticism has worked hard to reconcile "Render unto God that which belongs to God, and to Caesar that which belongs to Caesar."

As we have indicated elsewhere, these different ways of construing and perceiving the world are not only potential ways for human beings, but necessary for their fullest humanhood. A large and growing body of evidence suggests there is a need for humans to be-in-the-world in each of the four general ways demonstrated by the situation of the businessman. Indeed, it is beginning to seem more and more reasonable that much of our social (and personal) pathology may be related to a lack of ability on the part of many individuals to express these needs for being in acceptable and constructive ways.

This viewpoint that a human being needs more than one construction of the world—"realities"—to live and to achieve the fullest "humanhood" is one that is echoed by every serious mystical training tradition ("esoteric school") of the past and the present. The belief in the validity of this viewpoint has been gradually growing in acceptance in the social sciences in the past half-century. We have, however, progressed further than the mystical training schools and today are beginning to see that at least four different ways of being-in-the-world, constructing reality, are needed if the individual is to avoid becoming stunted in his development. Plutarch wrote:

If we traverse the world, it is possible to find cities without walls, without letters, without wealth, without coin, without schools or theatres: but a city without a temple, or that practices not worship, prayers and the like, no one has ever seen.[24]

Today we would add that no one ever saw a city where the inhabitants did not play, dream, deal effectively with objects according to the rules of the see-touch realm, or have ways of organizing reality so that there were no boundaries within it and everything was a dynamic One.

The four classes of ways of organizing reality that we tend today to see as universal and also as necessary for an individual to achieve the fullest potential are as follows:

The Sensory Reality. This is the everyday, "common sense" Western way of organizing reality. It is the way that our businessman used when he was at his desk and working. It is what the mystics tend to call "The Way of the Many." We all know the laws and Basic Limiting Principles of this organization of reality very well. It is essential to biological survival and is the way that must be used to cross a busy highway so as not to be run over by a truck. Its laws and entities are very similar to those of the see-touch realm of experience.

The Clairvoyant Reality. This is the way that the businessman was using when he was dancing and later when he was listening to music. In this way of organizing reality there are no boundaries, and nothing is separate from anything else. All things flow into each other and are part of a larger whole that makes up the entire cosmos. The individual relates to the whole as the single brushstroke relates to the entire painting or the single note to the symphony. The mystics have tended to call this "The Way of the One."

The Transpsychic Reality. This is the way of organizing reality that the businessman was using when he was praying, "Please don't let it be meningitis. " In this construction of what is, the individual is perceived as an entity that exists in its own right, but is also a part of the total One of the cosmos, so that no definite line of separation is possible. The analogy of the wave and the ocean is often used. Similarly, we see references to the arm and the body in the descriptions of this reality. It is the construction of reality in which intercessory prayer is perceived as effective. The individual is seen as separate enough to have wishes and desires, but connected enough so that it is possible to "urge" these on the great forces that make up the total cosmos.

The Mythic Reality. It was this construction of reality that the businessman was using when he was dreaming. It is the way reality is organized in play, myth, and sympathetic magic. In this way of organizing reality anything can be identical with anything else once they have been connected with each other spatially, temporally, or conceptually. The part is identical with the whole, the name with the thing, and the symbol with its object. Each can be treated as if it were the other. The world is full of all kinds of possible combinations and syntheses. The relationship between play and creativity has been remarked upon by a number of observers. This mode is necessary to keep us fresh and alive, curious and creative. Without it, "All work and no play makes Jack a dull boy," as the old saying goes—and there is truth in it. Without the ability to use this mode of being we become bored, and the sunset, our daily life, and even sexual activity becomes a dull affair.

We are aware that this has been a very brief summing up of the ways of organizing reality, but a further discussion here would be a digression. We refer those of you who wish further material on this subject to reference #23. We should make two further points about the different "realities." First, we have described them as metaphysical systems with laws and definitions basic to each. This is a valid approach and the only way to describe them clearly. However, from the experiential point of view, these "realities" must be seen as states of consciousness. When the individual is perceiving the world and reacting to it as if his set of Basic Limiting Principles is the *real* one, that individual is in an "altered" state of consciousness—altered from the everyday waking Western state of consciousness.

Second, these various states of consciousness deal with the same phenomena but with different conceptions of how reality works and what are its laws and goals—different definitions of space, time, causality, how things interact and what is the nature of a "thing." But they are the same phenomena. Whatever it is we take from whatever is "there" we carve up in different ways, just as Monet, Leger, Picasso, and Wyeth will each find a different painting in the same model. In his Introduction to his *Philosophic Investigations* Wittgenstein wrote that the book was not a coherent answer to problems, but an "album of sketches" from which we may be able to sense a picture of landscape. Experience is so complex, he wrote, that no one angle of vision would be sufficient.

To sum up, social scientists have become aware that their subjects—individuals and cultures—use a number of different conceptions of reality at different times. They have studied these and analyzed the rules and laws used in them and under what conditions each is used. Their basic attitude, with a few exceptions, has been that there is only one "correct" conception of reality, the Western, waking "common sense" description of reality: This is essentially the descriptive picture the physicist applies to the see-touch realm and generally corresponds to what is usually called "classical" physics.

Due to this belief in the correctness of this description of reality and the general belief in our culture that the entire universe is run on the same principles, social scientists have usually attempted to deal with the data in their field of research as if the data fit the see-touch picture of reality. Stating this in another way, the social scientists decided in advance of gathering and interpreting their data what basic laws and what kind of laws would make the data coherent and lawful.

When we examine the data in which the social scientist is interested, we find that they fall into two realms: that of molar behavior and that of inner experience. In these realms one of the most crucial and basic laws of the see-touch realm—the Second Law of Thermodynamics—does not operate. A new observable, "purpose," appears. In addition, definitions of such factors as space and time must also be different in order to make the data relate lawfully to each other. Other important differences also exist. It is plain that a different construction of reality is needed in these realms in order to deal with them scientifically. This is similar to the discovery the physicist has long since made about the microcosm and about the realm in which things are too large or whizzing by too fast for our sensory systems to perceive—even theoretically.

Major progress was made in physics in many areas only when the machine model was abandoned as a system of explanation—a description of reality—in these areas. When this was carried further and it was realized that the general description of reality used in the sensory realm was inapplicable in dealing with many different kinds of data, even greater progress resulted. We are suggesting here that much greater progress will be made in the social sciences when the students of this field abandon the assumption that there is only one set of principles on which the entire cosmos operates, and that this can be most accurately represented by the principles and laws of the see-touch realm, essentially the machine model of reality.

13 The Domains of Art

THE DOMAINS OF ART have been of astoundingly little interest to the scientist *qua* scientist. (As a person, he may be deeply involved in an art form. As a scientist he rarely employs his tools in this field.) One could read a dozen introductory textbooks in psychology and sociology and hardly find a hint that human beings in every culture and clime of which we know produced art and music and behaved as if they were important. Why it is, we must ask, with so much respect paid to a da Vinci and a Mozart, with museums in every city, with music from Beethoven to rock constantly being played, that the social scientist generally acts as if this were trivial or nonexistent behavior?

One answer immediately springs to mind in terms of one of our theses. It seems hopeless to attempt to "reduce" the work of the artist and the musician to the laws of the see-touch realm. First and foremost this work appears to be essentially nonquantitative. Plato certainly wrote of the relationships of mathematics and music, and it has been said of Bach that he "spoke in mathematics to God," but we feel deeply that we cannot capture the *Ninth Symphony* in an equation or explain the effects a Goya portrait has on us with a mechanical model. Feeling that we could not find the same laws operating in the domain of the artist and musician that we have found in the see-touch world, social scientists have largely ignored these areas of experience. In this they have followed a path similar to that of the Behaviorist psychologists who, when they could not fit the workings of consciousness into the laws of science found in the sensory realm, acted as if consciousness did not exist.

[169]

The physicist Max Born wrote:

The principle of objectivity can, I think, be applied to every human experience, but is often quite out of place. For instance, what is a fugue by Bach?
Is it the invariant cross-section, or the common content of all printed or
written copies, phone records, sound waves at performances, etc., of this
piece of music? As a lover of music, I must say No! that is not what I mean
by a fugue. It is something of another sphere where other notions of it
apply, and the essence of it is not "notions" at all, but the immediate impact
on my soul and the beauty and greatness. In cases like this the idea of scientific objectivity is obviously inadequate, almost absurd.[1]

In trying to bring the approach of Domain Theory to bear upon
the world of art, we should not expect our task to be easy or simple.
As we try to define this field, we are faced with a vast range of observables: material used and technique with which it was worked;
the artist's vision and his striving to accomplish that vision; the vary-
ing effect of a work of art on different beholders; the economic value
of a Rembrandt painting; different "schools" of art and music; and a
host more. Further, the responses to our questions as to the meaning
of "space," "time," "state," and "observer" needed to make the data
lawful are often very different among the various questions that we
can legitimately ask about "art." "Space" in Picasso's *Guernica* is not
the same as space on the walls of a museum, nor is it the same as
space inside a cathedral or space in classical ballet. "Observation"
and "measurement" mean different things when we are observing
the effect Byron Janis playing Chopin has upon us, when we are
counting the number of notes in the composition, or when we are
searching for what Chopin was trying to do when he wrote it.

After observing only a few of these differences, it becomes obvious
that there is no domain of experience into which we can place the
field of "art." It will require a number of them. In beginning the
analysis of this field from our viewpoint, we will describe four domains that are necessary to accommodate the field of art. (There may
well be more, but we not attempting here to present a finished science—whatever that strange phase might mean. What we are attempting to do is to describe the method 20th-century science has
developed and to show how it can be applied in a number of domains
of experience.)

The four realms into which we shall tentatively divide the field of
art are (1) The intent of the artist; (2) The responses of the beholders;

(3) The domain of man-made things; and (4) The domain of the medium (painting, music, poetry, sculpture, etc.).

We shall discuss the reasoning that leads us to this classification, but we should first note that each of these domains must be considered separately and that the observables, laws relating them, and the special meaning of guiding principles ("space," "time," etc.) in each must be compatible with each other and with the rest of our experience.

In terms of our earlier comments it should be clear from looking at the above listing that these four domains do not all belong to the same realms. The first two, the intent of the artist and the responses of the beholders, belong to the realm of the inner life, of consciousness. The third, the domain of man-made things, belongs to the sensory realm. From this viewpoint works of art are objects that can be seen and touched. The fourth domain, the domain of the medium, is perhaps the most surprising one. When we examine it closely, we shall find that each of the major artistic mediums is in a separate realm. Each requires a separate metaphysical system to make the data in it lawful. In poetry, painting, sculpture, and dance we receive quite different answers to our questions as to the meaning of the terms "space," "time," "state," and "observer." (There are, of course, similarities between the different realms of art. In every art form, for example, there are observables that the artists of every medium call "tensions" and their "resolutions." These belong to the "space" and "time" of the particular medium, as does the space of a painting or the space of a cathedral. In Euclidean space—the space of the sensory realm—there are no such things. Further, these observables are different in each medium.)

The Intent of the Artist. In 1911 the art historian Worringer[2] clearly pointed out (as Th. Lippe had done before him) that a theory of art must take as its starting point what the artist is trying to do. Any theory that concentrated on the artist's ability and used the assumption that he was trying to achieve *our* goals was bound to fail. For Worringer, what the artist willed to do (and therefore his world outlook) must be of primary importance.

Let us start our exploration with a remark made a few years ago by the pianist Artur Schnabel:

I never play a piece that is bigger than I am. It is too boring. There is some Chopin and some Liszt I could find the final way to play. So I don't play them. There is no point in it.[3]

The painter Rico LeBrun, in describing his own work, wrote:

To understand means to look. To *look*. No one can say precisely what the
aspects of nature mean to a man who is trying to find forms for his own vision. Sight, relentlessly in search of truth is a taskmaster.[4]

The artist Paul Klee wrote: "The artist does not reproduce the visible; rather he makes things visible."[5]

Picasso said: "I see for others."[6]

The artist attempts to find forms for his inner vision: to clarify,
sharpen, and express his own perceptions through giving specific
form to them.

The painter Fairfield Porter wrote: "The artist does not know
what he knows generally, he only knows what he knows specifically.
What he knows in general—or what can be known in general—only
becomes apparent after he has put it down."[7]

Goethe wrote: "The mind conquers by giving form to the indeter-
minate."

Works of art specify no immediate action or limited use. They are like gate-
ways, where the visitor can enter the space of the artist, or the time of the
poet, to experience whatever rich domain the artist has fashioned.[8]

The artistic symbol *qua* artistic (unlike the scientific) symbol negotiates insight, not reference; it does not rest upon conventions, but motivates and
dictates conventions.[9]

E. H. Gombrich wrote: "What the painter inquires into is not the
nature of the physical world, but the nature of our reactions to it."[10]

The artist sees himself and the rest of us human beings in the universe "as cats and dogs are in libraries."[11] He endeavors to learn to
read the books himself and then to show us what they contain. Each
book that is of interest to the artist says something new about our
inner life and offers new ways to organize and perceive reality.
Music, for example, carries information of this kind exclusively. It
carries no information about the "outer world" or instuctions on how
to deal with it. The composer or performing musician discovers parts
and aspects of his own inner world and communicates some of this to
his listeners. In the search for discovery new ways are invented, and
it is impossible to say where the discovery ends and invention begins.

Artists talk little about art theory. They are involved in striving to
permit reality to speak to them in new tones, to express these new
tones through a medium so they will know what they are, and to

communicate them to their audiences. As Santayana once remarked: "Art critics talk about theories of art. Artists talk about where you can buy good turpentine."

There is also a form of art produced with quite a different intent. At its best this includes decorative art, whose function is to make the world a more pleasant place to live in. It is certainly better and easier for most of us to be in a room in which the color scheme is coherent, and the colors blend, than one with just one color throughout and straight edges to everything. Background music and colors make it easier for us to relax and to be at ease in a world that, without them, would often be bare and unpleasant. Further, there are many people who have never developed their inner resources—and some who do not have the capacity to make this development. For these people being alone with their thoughts (or lack of them) can be quite distressing. For these the jingle and the pop song, providing an illusion of an inner life, is a great boon. And, indeed, for nearly all individuals there are times of fatigue when the detective story and the television situation comedy are wonderful ways of relaxing and resting: They take one away for a brief period from the realities of everyday life in the same way that the story teller in the bazaar entertains his audience.

Decorative art thus shades into the kind of art whose intent is simply to distract the individual; to help him "pass the time." It tries to render the observer unconscious of his thoughts or feelings while he is still technically awake. Whether called "jingles," "calendar art," or whatever, its goal is to reduce awareness by rhythm, line, color, or words. If we wish to consider an example of this type of art and the others we have been speaking of, we need only contrast the typical Western story and Cervantes' *Don Quixote.* The author of the first attempts to lower our awareness of the world while we are reading; the author of the second to heighten and deepen our knowledge of what it means to be human.

One of us once heard that superb lyricist, E. Y. Harburg, talking to a younger, very talented songwriter. Harburg explained (with many demonstrations on the piano) that the younger man was at a crossroads. He could either write songs that would make both the listener and the composer more aware of himself and help him remember the meaning of his existence, or he could write very popular songs without attempting to do this, songs that would almost inevitably have the opposite effect. Harburg felt strongly that the younger writer had the talent to follow either path. The younger man listened with respect, thought about it, and finally decided to follow the second path. He became well known, successful, and rich, and it is doubtful that any of his music will outlive its brief surges of popularity.

One important point that is often overlooked is the artist's use of talent to achieve something that physical scientists and even psychologists find difficult to accomplish. It is to convey in a manner that requires no operational definitions, no rules of correspondence, no measurements, the states of the artist's consciousness, his inner feelings. One good illustration of this is perhaps the last picture of Vincent van Gogh, *Wheatfield with Crows*. The picture was painted soon after his departure from the asylum at Saint Remy, a few months before he committed suicide in 1890. One description of this reads:

Signs of his grief—and his fears—abound in this turbulently emotional work. The sky is a deep, angry blue that overpowers the two clouds on the horizon. The foreground is uncertain—an ill-defined crossroad. A dirt path seen in part in the foregound runs blindly off both sides of the canvass; a grass track curves into the wheat field only to disappear at a dead end. The wheat itself rises like an angry sea to contend with the stormy sky. Crows flapping over the tumult swarm toward the viewer. Even the perspective contributes to this effect; the horizon rolls relentlessly forward. In this picture Van Gogh painted what he must have felt—that the world was closing in on him and his roads of escape were blocked, with the land rising up and the sky glowering down. Created in the artist's deepest anxiety, the painting nevertheless reveals Van Gogh's power, his expressive use of color and firm sense of composition.[12]

The Responses of the Beholders. A few years ago there was an exhibit of the work of *Les Fauves*, the impressionist painters who made such an impact on the Paris of their time by showing a new picture of reality, a new way of seeing what is all around us, that they were dubbed "The Wild Beasts." Deeply affected by the exhibit:

I realized that I could now see the world in a way that was new for me. I could look at a crowd of people, or at people and buildings and see them as if they were a Derain or a Seurat painting. I had gained something new and learned something about myself and my potential ways of organizing and perceiving the world.[13]

In ancient China, a painting was not exhibited, but unfurled before an art lover in a fitting state of grace; its function was to deepen and enhance his communion with nature.[14]

A little girl, after hearing Beethoven's *Ninth Symphony* for the first time, asked, "What must we do now?" Rilke, in his "Ode to an Archaic Apollo," wrote of the effect of seeing the statue for the first time. He ends with, "You must change your life."

Goethe wrote of art giving rise to feelings and understandings that were otherwise obscure or inchoate.[15]

Poetry does not move us to be just or unjust, in itself. It moves us to thoughts in whose light justice and injustice are seen with a fearful sharpness of outline.[16]

The artist produces articles of value. We are suggesting here one answer to the question of what that value is. The artist shows us new possibilities of ourselves and the world. It is for this reason that the first artist in a new school who gives us a new vision of the possibilities is regarded as a genius. Later artists of the same school, although their talents may be as great, are frequently seen as secondary figures.

Art also helps to state and reinforce the accepted cultural view of reality and how best to respond to it. This, too, is an important function. But it is not only in what we ordinarily mean by "new realities," "new constructions of the universe," that the artist aids us in our search. Of perhaps even greater importance is the fact that in each period we constantly search for new balances and syntheses of reason and emotion, individuality and relationship, appearance and essence, tension and relaxation, logic and intuition, doing and being, Apollonian and Dionysian ways of life. Within each world-picture the artist searches for these balances and presents the fruits of his searchings to us. In these areas also, he helps us try to remold ourselves and the world "nearer to heart's desire."

Some years ago one of us, as part of a group of graduate students, had the opportunity of observing an expert and loving child psychiatrist, Edith Meyer, working with a new patient. She had told the group that the eight-year-old boy who was coming in had a symptom that puzzled her. Repeatedly and obsessively he asked the question, "Is the grass green?" and no answer given so far seemed to satisfy him. Dr. Meyer had spoken with the parents but had not yet seen the child.

The students went into another room and watched through a one-way mirror. The child, who seemed rather tense, and his mother came in and were greeted by Dr. Meyer; after a few minutes of conversation, the mother went out to the waiting room. The boy and Dr.

Meyer talked for a little while and then, during a pause, and while both were standing and looking out the window, he turned to her and with a very serious expression and tone asked, "Is the grass green?"

She looked at him with what appeared to be complete attention for a very long minute and then replied, "Yes, I see things the same way that you do." The boy sighed deeply and seemed to relax all over. Later, in a follow-up meeting with the students, Dr. Meyer told them that he never asked the question again, and was much more relaxed, but that she was continuing to work with him.

Another aspect of this domain of art is illustrated by this story. This is that art can let us know that we are not alone with our perceptions, that others also see things as we do. We differ from others, and each of us is unique and therefore, to some degree, stands alone in the universe he constructs, and this can make for a great and sad loneliness. When we see a picture to which we respond with a "shock of recognition," with a feeling of, "Yes, that's how it is," when we find fictional characters with the same world-picture that we have, when we respond to a work of music with a feeling of "rightness" and a lessening of tension that we hardly knew we had, then the work of art is fulfilling the second purpose. A warm arm has reached out of time and distance to us and has let us know that we are not alone, that others live in the world and see it as we do. The artist has repeated to us the message of Plotinus: "None walks upon an alien earth." "Art," wrote Joseph Wood Krutch, "is successful only when we recognize ourselves in it, only when we are able to say, 'This is indeed what living feels like.' "[17]

Since the domain of the responses of the beholders is in the realm of consciousness, of inner life, the meaning of the terms "measurement," "state," and "causality" will not be those meanings necessary to make the data from the sensory reality lawful, but the meanings necessary to make the data from the realm of consciousness lawful. For example, although it is theoretically possible to obtain a precise numerical measurement of the effect of a work of art on an observer, and to use objective and quantifiable techniques, if the same technique is repeatedly used on the same observer, the results will be different each time. This is due to the fact that the observer is different each time he is exposed to the work of art. He is never the same twice

... even though the physical repetition may be exact as in recorded music, because the exact degree of familiarity with a passage affects the experience of it, and this factor can never be made permanent.[18]

This is a corollary of the fact that in the realm of molar behavior and of the inner life it is impossible to predict specific future events even though we can perceive, after the event occurred, that it was determined and that it had to happen exactly as it did (see chapter 7). If we found that we could get the same response to a test of the effect of a work of art on an observer each time we made it, we would be able precisely to predict events in the realms of molar behavior and the inner life. This would create a paradox—a contradiction between the laws governing different realms. Such a contradiction is not possible.

The fact that the "same" observer responds differently each time the same stimulus is presented is similar to the situation in the realm of the microcosm. Here also the "same" preparation will give different results each time it is tested. There are, however, quite different reasons behind the similarity of the facts in these two realms.

A science of this domain of the responses of the beholders, a science of "evocation," will have to consider the individual psychology of the observers and, because of this, their cultural backgrounds. For example, the Gothic cathedrals with their space of "unbridled activity ... meets the needs of Gothic religion and its striving for liberation."[19] To classical man, with his more Apollonian orientation, it would have far less meaning. Classical man had no wish to lose himself; Gothic architecture is designed for that purpose.

It may well be that we do not yet have the specific techniques to make these measurements (of the effect of a work of art upon the beholder). Our tests may not be precise enough. The answer to the question as to whether or not we already possess the necessary technical devices to make them must be tested by trying to use the relevant testing techniques, which we now possess. This has not yet been attempted. However, if we do not yet have the techniques, we do have the skills and knowledge to develop them if we wish. Once a science defines the technical devices it needs, these (unless they are theoretically impossible) are soon developed. This is certainly our experience in other fields.

The data from this domain all come from those situations in which the beholder is responding to the work of art and not approaching it in terms of description or classification. Rilke, in his *Letters to a Young Poet*, wrote:

Works of art are of an infinite loneliness and nothing to be so little reached as with criticism. Only love can grasp and hold and fairly judge them.

Freud once wrote: "The essence of analysis is surprise." Indeed, this is the essence of any growth or change. It is only when we let ourselves be open to experience, to being surprised, that we can experience anything new, anything we have not decided on in advance. It is only in this way that we can change. If I decide in advance what the experience should be, I cannot have a fresh experience. The beholder's response must be free and unplanned in order for the experience to be in this domain.

In the domain of the beholder's response a work of art, as Roger Fry has repeatedly pointed out, is "a glass and a transparency." It disappears into its own meaning as language disappears into what is being communciated. To the degree it does not disappear into itself, it is being treated as an object in the domain of man-made things.

The Domain of Man-Made Things. The study of this domain started with the descriptions of works of art in the biographies of Italian Renaissance artists. When we deal with works of art as objects that can be seen or touched, objects in the sensory realm, we can classify them in any way or ways we find convenient; by size, shape, or length; by form, as a sonata or a concerto; by school or century or geographical area of origin; or by economic value.

We must stress once more than no domain is more valid or less valid, more or less *real*, than any other. We mention this again at this place because so often those interested in art from the viewpoint of the domain of man-made things—as for example, those specialists who stress technical details such as the brushstrokes of a painting or the number of holes in ancient statues—deride (and are derided by) specialists in other domains of art. They will denigrate as "unscientific" those interested in the intent of the artist or the response of the beholder. Those specialists interested in the other domains of art will denigrate those working in the domain of man-made things as "pedantic." However, once we have accepted the concept of different domains of experience applying in the field of art, we can see not only the validity of each domain and the fact that different domains demand different methods of research, but also how the study of each one contributes to knowledge in others. The knowledge of the date of a painting can offer important clues as to the intent of the painter (in terms of the cultural orientations of his society), to the medium used (as in terms of the artistic inventions available at the time), to the cultural backgrounds that permit some beholders to respond and not others, and so forth. We shall return to this subject in our discussion of art and culture later in this chapter.

It is of interest here that specialists in the classification of works of art, have not yet been able to find an adequate language. We have used the language of biology for our descriptions of art in lieu of one devised for this domain. We speak of "the birth of a school," its "flowering," and its "decline and demise." We go on to talk about "offshoots of a school," the "maturity of a style" and its "fading," of a "school that died an untimely death," the artist who "branched off in his technique," the artist whose work "contained the seeds of a new revival," and of "the last gasp of Mannerism." We say such things as, ". . . the Gothic will to form exhausted itself and ran itself to death in this, the highest production of its energy."[20] We speak of "Roman Baroque" and the "Umbrian School" ". . . in a manner vaguely patterned upon biological classification by typology, morphology and distribution."[21]

Thus we have taken a specialized language from the see-touch realm and applied it to a domain quite other than the one for which it was formulated and to which it is applicable. In terms of the presently available knowledge about how the language we use to describe something shapes and limits our thinking about it, it seems little wonder that we have made so little progress with the classification of art; that this domain still awaits its Linnaeus.

From our viewpoint a "work of art" would have an artist's intent, a medium, and beholder's response. Without the first it is a beautiful object or scene—a sunset or a piece of driftwood. Without the third it is an "objet d'art," a work that is classified rather than responded to. From this viewpoint a *work of art* can change to an *objet d'art* as the culture changes so that it is no longer responded to. It can change back again as a culture learns again to respond to it.

Primitive and African art works were—for most Westerners— objets d'art to be classified, until a Picasso taught us how to respond to them. Then, for the same Westerners, they became works of art. "Objets d'art belong only to the period in which they were made: works of art belong to our own period as well."[22]

When considering works of art from various viewpoints such as size, shape, economic value, school, or tradition, we are considering

them as observables in the domain of man-made things. This is one of
the legitimate and necessary domains of works of art. It is a domain
in the sensory, the see-touch realm, and the Basic Limiting Principles
as well as the definitions of "space," "time," "causality," etc., of this
realm apply to it.

The Domain of the Medium. By the term "medium" we mean here
such specific arts as painting, sculpture, architecture, poetry, and the
dance. Let us be consistent and approach each of these with our
questions: What are the observables found here? What are the laws
relating them? What is the meaning of the terms "space," "time,"
"state," "causality," needed to make the data lawful? If we do this,
we find that we often obtain different answers in different media. If
the meaning of the basic terms is different in each, then we need different metaphysical systems to make the data lawful. We find to our
surprise that there is no *one* domain that will include these different
media, but each is in a separate domain and a separate realm. A
realm—as we have indicated before—is composed of a domain or
group of domains in which a single metaphysical system can be used
to make the data, the relationships between the observables lawful.
A domain needing a different metaphysical system is in a different
realm.

We do not plan to analyze in this chapter the different artistic
media from this point of view: We do not have the skill and training,
and this is not the place for such an intensive exploration. We shall,
instead, briefly examine one medium as an example—painting—and
ask one question: "What is the meaning of the term 'space' in a
painting?"

Space in painting (what the philosopher Suzanne Langer calls
"virtual space"[23]) is only present to one sense and has no connection
with other senses. Being only visual, it has no connection with the
space(s) in which we move and act (personal space and Euclidean
space, etc.). Its boundaries do not separate it from these other spaces
because any boundary that separates also connects, and there is simply no connection between these spaces. These different spaces are
equally real but unconnected. One does not even "interrupt" the
others. The created visual space is entirely self-contained even
though it is perceived to extend in all directions "behind" and beyond its boundaries.

Hildebrand the art theorist discusses the relationships of "pictorial
space" to "practical space." He shows that pictorial space is real

and *in* practical space, but does not interrupt it, does not make it discontinuous.

Let us imagine total space as a body of water in which we may sink certain vessels, and thus be able to define individual bodies of water without, however, destroying the idea of a continuous mass of water enveloping all.[24]

The "space" of pictorial art is shown most clearly in a motion picture, which is a complete visual experience, separate from the space around it, architectonically related to itself, extending off in all directions to infinity, opposite the eye and related to it directly.

Whatever the "objective," sensory-realm-size of a picture—however much it fills or fails to fill our visual field—it is a total visual field in and of itself. There is no relationship between the sensory-realm size of a picture and the extent to which it is a complete visual field; it is always the latter.

A simple and interesting experiment can demonstrate this. Imagine a painting concealed in such a way that we can only see it through a tube the same shape as the canvas. All we can see through the tube is the painting. There is nothing else in the visual field to compare it to. It is impossible from the visual experience alone to tell if we are looking at a miniature or a giant mural.

(Although the above comments apply to pictorial art, it is also true as a general principle of art. There is no relationship between sensory-realm space and size and virtual space and size. The Louvre houses the magnificent *Nike of Samothrace.* As you gaze at her from the bottom of the steps—she stands facing you at the first landing—she is obviously of the "correct" size; you would not dream of shrinking or expanding her size; it is "right," and the question does not come to mind. Then you notice a glass cabinet to the right of the steps. The arms of the statue were never found, but one of the hands was, and it is in the case. Looking at it, as it lies there unconnected with the statue, it appears oversize, bulbous, gross, revolting. You then realize that the entire statue is several times life size.)

The *attributes* of space in painting are also very different from the attributes of Euclidean space. The painter and art theorist Hans Hofmann wrote in this connection:

Depth, in a pictorial plastic sense, is not created by the arrangement of objects, one after another, towards a vanishing point . . . but on the contrary . . . by the creation of forces in the sense of *push* and *pull*. . . . The forces of *push* and *pull* function three-dimensionally without destroying other forces functioning two-dimensionally. To create the phenomenon of *push* and *pull* on a flat surface, one has to understand that by nature the picture plane reacts automatically in the opposite direction to the stimulus received. . . .[25]

We have been using pictorial space as an example. Each medium of art has its own definitions of space. Each is different from psychological space with its before-behind, far-near, up-down, left-right attributes. Each is also different from space in the microcosm, space in the sensory realm, or space in the macrocosm.

A brief digression from our analysis of "space" in pictorial art may be in order here. Let us look quickly (and, regrettably, superficially) at some of the other "guiding principles" we use to organize our perceptions and see how they apply to this domain. Time in the domain of the medium is not the time of the sensory range, the time of one-thing-after-another. It is the time in which myths exist, the time of everything-happening-all-at-once. Many things are going on at the same time; the whole picture is "happening" at once. The effect of the *Hammerklavier* or a Monet *Waterlilies* is the effect of the total work on the listener. This is its meaning. Any other analysis does not deal with it in its own terms, but in terms of another domain.

I once tried to suggest art's own time by comparing it to the time of saints when one prays to them. The faithful believe that a saint belongs to the present, which his eternal life bestows upon him, and during which prayer takes place. He also belongs to historical time since he has a biography . . . Finally he belongs to chronological time, to the duration of the living. Admiration "actualizes" a work of art just as prayer actualizes a saint.[26]

While we are only at the earliest beginning of our analysis of the domains of the medium, one other aspect, comes immediately to mind. Objects in the see-touch realm may be reduced to their components, studied this way, understood in their functioning more deeply, and reassembled. When reassembled, they work as well (or sometimes better, if they have been cleaned in the process!) as they did before. Things in the domain of living organisms can be reduced to their components, and these can be studied. We can then understand more completely how these components work and how they relate to each other. However, from this process we do not appear to be able to learn what "life" is, and we certainly cannot reassemble the components and have them again functioning and alive. Works of art can be reduced to their components as we count the brushstrokes in a painting, analyze the interplay of form and color, see what happens when we change the medium from bronze to marble, or count the number of times Shakespeare used each word in his lexicon. But these tell us nothing of the total effect of the work of art and seem quite irrelevant to the questions of this domain. We can reassemble the work of art after disassembling it, and if done skillfully, it seems as good as new!

In exploring each new domain, we must be conscious that we make many assumptions of which we are very little aware. These are rules and expectations we project on reality, unconscious or barely conscious that we have made them.

These generalizations from our experience in the sensory realm may or may not be valid in other realms, but we cannot know until we have become consciously aware of them and tested them. One of these, for example, is the concept that opposites balance out to a central point of truth. A pound of water at 40 degrees and a pound of water at 80 degrees, if mixed, add up to two pounds of water at 60 degrees. Two ounces of weight on one pan of a good balance scale and two ounces on the other do bring the pointer to the exact center. However, in realms in which the observables are not quantifiable in principle the concept is meaningless and serves only to block and confuse our thinking. If we deal with factors such as hope and fear, form and content, Dionysian and Apollonian, pain and pleasure, nature and nurture, instinct and intellect, rights of the individual and rights of the state, freedom and tradition, the concept that opposites automatically balance out to a precise middle point is simply not applicable. With nonquantifiable observables and nonlinear systems there is simply no such "middle." Mental processes do not obey linear laws since there are no numbers which can be meaningfully attached to them. One cannot add two units of joy and two units of grief and expect to get two (or four) units of serenity. In many realms "the concept of polarity, intriguing though it be, is really an unfortunate metaphor whereby a logical muddle is raised to the dignity of a fundamental principle."[27] In the domains of art we must beware of this concept being considered applicable until it has been examined for meaningfulness.

This very brief discussion of space and other guiding principles in painting will serve as an example of how the definitions of basic terms such as "space," "time," and "state" are different in different artistic media. Comparing the meaning of space or time in painting, music, and the dance will make it clear that they are different to such

a degree that different metaphysical systems are necessary to make
the data lawful. This is why we come to the conclusion that each artistic medium is a different realm of experience.

Art and Culture

The artist's search for meaning, values, and organization of the
cosmos is not chaotic or random, in each period of the development
of a culture it is limited in its possibilities and regulated by several
factors. First, just like scientists, artists are limited by the technical
methods they have available. Scientists could not study bacteria before the invention of the microscope, and artists are similarly restricted. Painters in Western culture today have many more artistic
inventions available to them than had the painters of the Renaissance, and therefore they have many more possibilities available for
shaping the canvas. The effect of this factor on art was described by
Nietzsche. The artist's goal, he wrote, is to paint:

> "All nature faithfully"—but by what feint
> Can Nature be subdued to Arts' constraint?
> Her smallest fragment is still infinite:
> And so he paints what he likes in it,
> And what does he like? He likes what he can paint!

The possibilities open to artists are also limited by the cultural
viewpoint within which they live. Each culture makes certain approaches to the infinite possible. It makes other approaches impossible or incomprehensible. A cartoon published several years ago
showed a studio of a Renaissance painter. A large number of typical
paintings of the period stood about and hung on the walls. In a cor-
ner was the famed Mondrian painting of 1921, *Composition with
Red, Yellow and Blue*. The painter was explaining it to a friend. "Oh,
that's just something I tried which didn't work out."

Of those attempts at the organization of reality made by its artists
each culture selects some as "successful" and rejects others. What
the culture selects then helps shape that culture.

The deeply reciprocal nature of the culture and the constructions
of reality used, the constant feedback and corrections between "nature" and "consciousness," the "epistemological feedback," can be
seen perhaps most clearly in this relationship between artists and society. Out of the variety of coherent possibilities that exist within the
limits of his cultural world-picture and the artistic inventions known

to him, the artist chooses a construction of reality and writes, composes or paints within it. The society chooses which of its artists to pay attention to, and then the artist's conception becomes a factor in shaping the society. The pre-classical Greek society chose Homer as its major artist. Homer then had a profound influence and helped build classical Greek society. We do not yet know enough of the factors involved that make a society choose one or another artistic conception. Once one is chosen, however, it appears to be foretelling the future of that society because it is helping to form it.

André Malraux has pointed out that, "We do not mind a Rembrandt looking modern, but resent a modern picture looking like a Rembrandt."[28] One reason for this springs from our previous statements. When an artist tries to paint a picture in a world-view in which he does not live—as a modern artist might try to paint a medieval religious picture—he is not exploring his inner landscape and trying to expand it. He is, rather, painting an outer landscape and describing it. It therefore emerges as a false picture. He is following the method of the scientist, the method adapted to the realm of science, not that of art. The method of science is to search and establish perceptual reality—what is perceived as *outside* of our inner experience—and to describe it so that we can perceive something new and then change. The method of art is to change our inner experience so that we then perceive the perceptual world (and our inner experience) differently.

The poet describes his total experience. A poem is an expression of and a way of ordering his experience. A scientist tries to describe his perceptual experience only.[29]

A medieval painting portrays a world in which all major questions are theological ones: a world in which Christ was the Perfect Man and all human endeavor is a striving toward and falling short of emulation. This is no longer a world in which we live. Almost anyone who today tries to paint the world of Giotto and Fra Angelico, of Dante and Nicholas of Cusa, is painting a world false to him and his paintings ring hollow.

When that rare exception, Dali, successfully paints a *Crucifixion* or *Last Supper* using modern techniques to explore an old vision of reality, we know that, at least during the time he was painting, he was perceiving and reacting to this world-picture. He was showing us, with the use of all the aesthetic inventions available to him today, what he felt at the time and, therefore, what we could also perhaps feel and see. He was living in this world-view and not describing it any more than a man who has taken LSD and who is still under its influence is *describing* an altered state of consciousness. Dali was *in* this world-view, and therefore his painting "worked."

If we try to re-create Gothic architecture today, we do it without being part of the original spiritual search. The Gothic thus emerges looking as if it had been designed by a computer, not by the human relationship with God.[30] If we do not comprehend the nature of the universe of Gothic architects, of their striving to bring man into a state of consciousness in which he directly relates to God, then we do not create this architecture in its own terms, but rather in terms of present world-views. One is then like the "naive" historians described by Nietzsche who ". . . gauge all bygone opinions and deeds by the generally prevailing opinions of the present and call this 'objectivity'. . ." and see their task ". . . to fit the past into the triviality of the present."

During the medieval period the chief task of art was to create the sacred world, to reorganize the observer's perception so that he was aware that the Passion was *now*. In this society the artist was controlled. As late as 1573 Veronese was reproved by the Inquisition for including aspects of the sensory world in a sacred painting. As the Renaissance developed, the artists gradually discovered the possibility of a great many different worlds. ("Painting is a form of poetry made to be seen," wrote Leonardo.) Artists could now search for all the worlds their imaginations could reach; all the visions of reality they could construe. Religious control over the arts and sciences had largely collapsed, and both expanded their explorations.

Roger Fry has discussed in detail the parallel between the scientific thrust of the Renaissance (to understand the world) and the major thrust of the artists of the period (to reproduce it exactly). For five centuries Western art was largely concerned with this problem ". . . either by an accurate imitation of an actual scene or by constructing a picture according to those optical laws to which our vision inevitably conforms."[31]

It seems far from coincidental that at the same time (the beginning of the Renaissance) at which the cultural orientation turned its efforts to the understanding and control of nature, art began its five-century attempt to portray a nature exactly as it appeared to the eye, and Giotto and his followers introduced a new idea—that nature was beautiful. Neither the classical Greek nor the medieval man would ever have thought of such an idea.[32] It seems legitimate to say that things do not appear beautiful to us until the artist has taught us to see this quality in them. It thus teaches us another way to arrange our categories of what we perceive.

It is doubtful if anybody found anything but dinginess and "ugliness" in the mist and fog of the Thames-side London before Whistler ... invested the murky London atmosphere with permanent beauty.[33]

There is a deep truth behind LaRochefoucauld's comment that if it were not for poetry, very few people would ever fall in love. The converse is ... also true. If a few people had not fallen in love, no one would have ever written poetry.[34]

In the latter part of the 19th century:

The aim of five centuries of European effort is openly abandoned. The actual appearance of the visible world is no longer of primary importance. The artist seeks something underneath appearances ...[35]

Picasso: "I paint as I think, not as I see."

Since artists search for new ways to organize reality, they are always a threat to the stability of a society. Those who envision a particular form of state or Utopia which they feel should remain unchanging, view the artist as a danger. From Plato's *Republic* to Stalinist Russia, from primitive societies to the "Thousand-Year Reich," those in power recognize that if things are to remain constant, the artist must be controlled or expelled. A new experience, a new vision of reality, can (and often does) give individuals new goals, hopes, and ambitions. Those who have power and wish to keep it must therefore view the artist with suspicion. Art—except in a rigid society that firmly controls it—implies and leads to independence of search, and if the search is successful, then others may begin to wish to change or to leave the hive.

The painter Constable wrote in his notebooks:

Painting is a science and should be pursued as an inquiry into the laws of nature. Why, then, may not landscape painting be considered a branch of natural philosophy, of which pictures are but experiments?[36]

The fact that art is a search for new ways of organizing reality is shown, perhaps, by the fact that there is a marked tendency for societies to give the same amount of freedom to art as they give to science; by and large both are controlled to approximately the same degree.

Art and science at any period in a culture have certain conventions and sacred cows. Those who violate them today in America will not have their pictures hung in galleries or see their papers accepted by scientific journals. There are other cultures with more rigid controls than present-day Western society. In these cultures violation of the conventions and rules of science or art will get one into a prison camp, mental hospital, or sometimes just thrown over a cliff.

Since each new development in art (as in science, as we noted in our discussions of reductionism) implies change in the constructions of reality available to us, we should expect that each new development would modify our perceptions of those that preceded it. And, indeed, that is exactly what we find.

After Van Gogh, Rembrandt has never been quite the same as he was after Delacroix. (Nor has Newton been the same since Einstein.) Each genius that breaks with the past deflects, as it were, the whole range of earlier forms.[37]

We have indicated earlier that the art and science of a period, the two main thrusts of our developing change of our ways of organizing reality, tend to move forward in the parallel ways, sometimes one appearing first as the spearhead of the new, sometimes the other. A change in artistic comprehension of reality may herald a change in the scientific world-picture, or vice versa. At the same time the tight enclaves of the medieval world were opening up to Renaissance; that Bruno was showing the implications of the concept of infinity; that science, no longer limited to theological problems, was being expanded in all directions by Galileo and others, art also aided and was aided by the new attitudes. For example:

In all the paintings known to Leonardo's world . . . (and in all those previous to him we have since discovered) . . . painters had always composed *in terms of outlines*. It was by blurring outlines, prolonging the boundaries of objects into distances quite other than the abstract perspective of his predecessors . . . it was by merging all things seen into a background suffused into various tones of blue that Leonardo . . . invented a way of rendering space such as Europe had never known before. No longer a mere neutral environment for bodies, his space enveloped figures and observers alike in the vast recession and opened vistas on infinity.[38]

As we look at the artistic inventions of Leonardo, we can see Newton in the wings ready to translate these changes in perception into changes in the organization of the outside environment. The extension of constructs hinted at in the above quotation clearly reminds us of Newton's idea of universal gravitation.

There is nothing—and this is hard for us to realize—that represents a "neutral" style, what the object "looks like" before the style is added. There is no purely *objective* way to perceive something, to perceive it as it is before consciousness shapes it into something we can perceive. What looks objective to us is what we construct and are accustomed to see. The art theorist Sir Herbert Read wrote:

We do not always realize that the theory of perspective developed in the fifteenth century is a scientific convention; it is merely one way of representing space and has no absolute validity.[39]

Extensive experiments by Thouless[40] showed that perspective is the way that we have been taught to perceive on a plane surface and not the way that objects in three-dimensional space appear to the eye. Santayana told of how:

He [Bertrand Russell] seemed one day astonished and horrified when I said that the image of the sun as a luminous disc, sometimes (if you squint) with rays around it, is as fictitious and imaginary as the idea of Phoebus Apollo with his golden hair and arrows. The senses are poets.[41]

This is amplified in our discussion of the establishment of physical reality in chapters 3, 4, and 6.

Art introduces us to new ways of perceiving and reacting to the world. The artist knows that there is no one *correct* way to perceive. He searches for new ways so that we may see the world differently. He seeks new and different views of reality, and when he is successful in his search, the culture learns to perceive with his new view. When Picasso was told that his portrait of Gertrude Stein did not look like her, he replied, "Don't worry. It will." Similarly, if Ibsen had been told that women did not behave like his heroines, he might well have replied, "Don't worry. They will," and seventy-five years later he would have been proven to be correct. Rudolph Arnheim, one of the rare psychologists who has seriously studied aesthetics, has remarked that the Sears, Roebuck catalog of the year 2000 may well have illustrations that resemble paintings by Braque and Miro because that will be how people will then be seeing the world, and these artists will be called "realistic."[42]

Perhaps the function of the artist and composer can be more clearly seen if we compare their work to that of the historian, who "... communicates a pattern which was invisible to his subjects when they lived it, and unknown to his contemporaries before he shaped it."[43]

We have pointed out that not all conceptions of reality are valid.

There are definite rules for establishing sensory reality. The "clay" of reality can only be manipulated into some shapes. It will not hold others. This is also true in art. There are limits to what the artist can do, and if he goes beyond them, his work loses its validity. The painter Rico LeBrun wrote of one aspect of this: "There is a point beyond which the human image refuses to play ball. Its structure has a terrible lack of acquiescence toward the pun and the decorative."[44]

In all realms the model of reality that has been chosen must be used consistently. Otherwise, whatever is being attempted fails of its goal as alchemy failed, due to the use of a mixture of two constructions of reality. When the premises behind alchemy were sorted out, and the use of the mythic mode-of-being removed and the sensory mode left, chemistry—which succeeds in its goals—replaced alchemy. This need is present in art as well as in science. The theatrical production in which the director has not been clear with himself as to what is the "level" he is using in his approach to the audience, is a failure. The painter Fairfield Porter wrote, "Reality cannot be faked; unless it is total, it fails to convince . . . It fails when it exists in the detail and not in the whole surface."[45]

Artistic inventions alter the sensibility of mankind. They all emerge from, and return to human perceptions, unlike useful inventions which are keyed to the physical and biological environment. Useful inventions alter mankind only indirectly by altering his environment; aesthetic inventions enlarge human awareness directly with new ways of experiencing the universe rather than with new objective interpretations.[46]

Aesthetic inventions are focused upon individual awareness; they only expand the range of human perception . . .

Human sensibility is the only channel to the universe. In the capacity of that, these channels can be increased, knowledge of the universe will expand accordingly.[47]

The media of art are not adapted to communicating the specific kinds of information about the world as well as verbal language or mathematics. Nevertheless, they are ideally suited to communication. This is what the composer Felix Mendelssohn meant when he wrote:

The meaning of music lies not in the fact that it is too vague for words, but that it is too precise for words.[48]

The tonal structure we call "music" bears a close logical similarity to the forms of human feeling—forms of growth and attenuation, flowing and slowing, speed, arrest, terrific excitement, calm or subtle activation and dreamy lapses—not joy and sorrow perhaps, but the poignancy of either and both—the greatness and brevity and external passing of everything vitally felt. Such is the pattern, or logical form, of sentience; and the pattern of music is the same form worked out in pure, measured sound and silence. Music is a formal analogue of emotive life.[49]

Music is precise in a realm in which verbal language is vague—in the realm of the inner life. We have shown earlier (see chapter 11) that our language for this realm is a series of metaphors (*hot* rage, *iron* determination, etc.). This is the best we have learned to do with pure verbal language. Painting, sculpture, dance, are other methods of communication in areas in which verbal language is not effective.

The types of communication the various art media are adapted to do not cover information about the outside world, but nonverbal insights and comprehensions of ourselves and our potentialities. As we develop and garden these, our perception of the outside world and our relations with it also change and develop.

The Domain of Man-Made Objects is in the sensory realm and must be treated according to the guiding principles governing this organization of experience. With this exception, however, the domains of art (and we would also include here those of philosophy and religion) are not sociological or anthropological investigations into the "what is" of human behavior, but a form of discourse into the meaning of value judgments and their implications. They do not present "techniques," but general approaches to ways of being with ourselves, each other, and the cosmos. Without this discourse we have only techniques with no beliefs to support them—except that whatever works is its own justification. The value and importance of art, philosophy, and religion are not—as in science—that they reach generally agreed-upon conclusions, but that they continue to expand and deepen the possibilities of our being, to make us more human rather than more machinelike. The goals of these domains do not aim toward agreement and conformity but toward diversity, communication, and choice; toward individuality and uniqueness; toward modes of feeling, thought, and behavior that can be tested individually only. These help us to become richer in different ways rather than tending toward the type of agreement on hypothesis and experiment we quite correctly find in the chemistry journals. In each realm of domains it is not only the observables and their relationships that must be treated in their own terms, but also the general guiding principles, *including the goals and purposes of organizing our experience in this particular way.*

One factor in the "success" (at least in terms of their long-continued survival) of such religions as Catholicism and Judaism is their basic insistence on the concept that they enriched and deepened human life no matter what were the social, economic, or biological conditions. This is seen, perhaps, most clearly in Judaism, which does not even allow questions about an afterlife to be asked. This religion promises nothing (as far as the sensory realm is concerned) in this world—see the story of Job, for example—or the "next"; it does not even promise that there is a "next." The "purpose of artistic and religious experience is to enrich life."[50]

The great art theorist Bernard Berenson wrote of the goal and purpose of serious art:

Our world may be nothing but the order that we . . . are shaping into a cosmos. The more we refine and perfect ourselves as instruments, the better the cosmos that we are winning out of choas will be.[51]

Beauty

The general aim of ordinary science, of the experimental and theoretical investigations that establish the constructs of physical, chemical, and biological reality, is said to be the attainment of truth. Truth is not a simple idea, however; hence we began our discourse on ordinary science with searching comments on the meaning of truth. The most general aim of art is the attainment and then the enjoyment of beauty. Beauty, like truth, is not uniquely definable. It seems appropriate, then, to take a closer look here at the meaning and the origin of beauty. One of the questions we wish to address concerns Somerset Maugham's statement that beauty is in the eye of the beholder and, presumably, in the ear of the listener.

We have found that there is no simple, universal definition of truth and have called attention to the differences between logical, linguistic or semantic and scientific truth. The latter is never fully at hand; we called it asymptotic, a goal always sought and never completely grasped. The implication is that truth may change in the process of being sought. We shall find the same to be the case for beauty. But neither truth nor beauty, taken by themselves, define the detailed realms of the different sciences and the different arts. This is done by the method we have outlined, which involves first the discovery of meaningful and methodologically promising observables and their combination into laws that can be shown to hold in the area under study.

Another feature that needs recording has to do with the concepts of compatibility and transcendence (discussed in our analysis of reducibility, chapter 8). There we found that one domain of experience or reality has observables compatible with, yet some of them transcending and therefore inexplicable in terms of the observables in some other domain. The same thing is true for the domains of art.

We may call attention first to features of art—i.e., observables related to beauty—that have their origin in science, then to others that transcend them, which arise in the pursuit of beauty beyond the areas where it makes contact with science.

Our treatment of physical reality placed emphasis on the guiding principle of simplicity—i.e., simplicity of scientific laws. One aspect of simplicity is called invariance, a term we hasten to explain. (In technical jargon a whole branch of mathematics called group theory deals with it, but we shall describe it in nonmathematical language.)

Invariance means precisely what it normally implies: something remains the same when certain specifiable changes in its surroundings occur. Thus we speak of a watch as being reliable when we mean that its running is "invariant" even when changes occur in temperature, in humidity, and in the motion of the wrist to which it is attached. It is *not* invariant under a hammer blow. A tune written in a major scale is invariant with respect to transposition to all other major scales, but not to a minor scale. Thus, in dealing with the property of invariance of an object, one must specify the object as well as the permitted changes. The object may be concrete, like the watch; it may be abstract, like the tune; in many important instances it may be a law or an equation, in which case the changes are called transformations. But these are professional matters that need not concern us here.

The ancients regarded the circle as the most perfect figure. The reason is this: If it is turned about its center in its own plane through *any* angle, its appearance remains unchanged. It is invariant with respect to all rotations. Or consider a regular polygon. A square is invariant, in the same sense, with respect to rotations about 90 degrees and all multiples thereof, a hexagon with respect to rotations about 60 degrees and its multiples, and so on. Artists call this particular kind of invariance "symmetry," and they regard it as an elementary aspect, an observable of beauty. The beauty of a snowflake, of a flower, of a crystal resides in its symmetry, in its invariance with respect to rotations.

There are other kinds of symmetry: Bilateral symmetry is invariant with respect to mirror reflections, something present in nature as well as in art. The aesthetic appeal of a regular pattern exhibited in wallpaper decorations is another kind of invariance, the permitted changes being "space displacements" through specific distances: The wallpaper pattern is unchanged if the paper is moved (in the proper direction) through a distance equal to the space between its stripes or figures.

Another aesthetic element, this time in music, is invariance with respect to "time displacement." A regular sequence of sounds is more agreeable than noise. The reason is that, if the sequence were advanced or retarded in time by an interval equal to that between the sounds, the change could not be detected. We call this feature, this kind of invariance, rhythm, and rhythm is an important element of music.

The examples clearly bear out the claim that some, perhaps rather primitive, ingredients of art are compatible with, and may be said to have their origin in, science. Space considerations forbid the discussion of numerous others, such as the harmony of musical chords, the peculiar appeal of major and minor scales, many of the features of classical music which at one time found their expression in rigid rules taught in courses on the theory of music.

The assignment of beauty to scientific theories, or even mathematical equations, though difficult to define, is not infrequent in current discussions. Paul Dirac, whose creative genius explained the existence of the positron, liked to speak of beautiful equations. On one occasion he presented a paper that worked out the implications of what he called a beautiful equation. Its analysis led to a result that prompted him to suggest that a specific new *on*, not yet discovered, should exist within the rapidly growing set of elementary particles. Afterward, during lunch, one of us asked him what he meant by a beautiful equation. His reply: "I am not a philosopher, but you know perfectly well what I mean by it." He refused to define it in formal terms. Then I asked: "Do you expect your new particle to be discovered?" He answered with a hopeful yes. Since this conversation took place fairly recently, the end of the story cannot yet be told.

Art goes far beyond the elements it shares with science, transcends them in many ways and creates its own peculiar realm of experience, its own observables. Conversely, many features of science cannot be incorporated into art. In this context we recall an incident that took place about thirty years ago.

The composer and theorist Paul Hindemith, who was at that time a professor at Yale, told one of the authors that he wished to compose a piece of music based on a very fundamental set of frequencies occurring naturally in atomic physics. Upon brief reflection we discussed, first in qualitative terms, the spectra of simple atoms and the range of their frequencies. Because the simplest atom is hydrogen and its most fundamental and simplest series is the so-called Lyman series, whose lines follow an attractive mathematical law involving all integers from 1 to infinity, I mentioned it as perhaps an attractive model but cautioned him about the diminishing intervals between lines at the series limit. The optical frequencies are, of course, much higher than acoustic ones but can be reduced to the acoustical range through multiplication by a constant factor. Hindemith asked that I send him the results of this conversion. A few days later he phoned, telling me with some regret that the physically basic frequencies were not suitable for any sort of acoustic transcription.

As in other instances previously cited, science leads into but does not exhaust the realm of the arts. Much of modern music and impressionism and surrealism in painting cannot be captured in the categories of chapters 4, 5, and 6.

14 The Worlds of Color of Newton and Goethe: Two Domains of Reality

THE SAME DOMAIN of sensory experience, when analyzed in a manner different from the one we have outlined in detail, can give rise to a view of reality other than the one to which we are accustomed. But this view will be equally useful for different purposes, in this case the aims of the poet and the painter.

Newton's theory of optics and Goethe's (we base this discourse on the complete edition of 1808) have rarely been viewed or analyzed as different forms of human experience, as different realms of reality. Most scientists think of Newton's theory as being right and of Goethe's as wrong. Artists in general agree with this judgment, but in veneration of Goethe's genius they declare his approach to the problem of color more vivid, more personal, more genuine as an immediate reaction to nature. Everyone who has been in the museum of Weimar and seen the display of colored plates with Goethe's words describing their direct sensory impression, their release of sentiments in the observing mind, will admire the charm of his interpretation. To be sure, he was unable to extend his constructs and his feelings toward a logical prediction of the many intricate sensory facts conjectured and observed by Newton. His reasoning had no practical effect on the construction of telescopes or the technique of photography. Yet for some strange reason his work on optics cannot be forgotten and is rarely regarded as an error of reasoning.

The simple fact is that Goethe did not use the method by which the scientist constructs reality, the method we explained in chapters 3, 4, 5, and 6. In a certain vague sense he proceeded from the P-plane to the right, entering into his mind rather than the world of external constructs, thereby creating a reality of a highly personal sort that, when verbally explained, seemed to resonate with other minds.

Many see this as a poet's prerogative or license. We cannot, however, agree with this view. Goethe was not unaccustomed to scientific investigation in the sensory world. Prior to his theory of color he wrote "Morphology of Plants," and even biologists credit him with having anticipated the idea of organic evolution half a century prior to Darwin. In Goethe's works many pages are devoted to an attempt at proving Newton wrong. Newton had at that time been dead for over fifty years and could not defend himself. In spite of this few readers fail to be impressed by Newton's very detailed arguments. There seems to be only one reasonable assessment of this notorious situation: Newton and Goethe presented alternate realities, one useful to the physicist, the other vaguely descriptive of any person's feelings but cherished by many artists as their proper domain.

The question of which of these realities is the correct one is unimportant. Truth, as we have seen, is an elusive concept. From our present point of view both interpretations contain errors. Newton's can be easily discovered because his theory is analytically explicit, while Goethe's is vague. Let us deal with Newton first.

As originator of the modern science of mechanics Newton accounts for the straight-line propagation of a light beam by attributing to it a corpuscular nature, assuming that it consists of rapidly moving particles, or corpuscles, shot out by a light source. They can be reflected by liquid and solid substances, and the law of reflection is the same as it is for perfectly elastic billiard balls. Refraction is accounted for by attributing different speeds to the particles when they travel in different media, so that they are slowed down or accelerated at the surface between two different transparent substances. Thus, when a light corpuscle enters a body of water from air and is, according to common observation, deflected away from the surface of entry, this could be explained by assuming that the particle was attracted by the denser medium. (This, incidentally, turned out to be erroneous, for when Foucault at a much later date—1850—succeeded in measuring the speed of light in different media, it was found to be slower in water than in air, whereas the attraction of the particles should have made it faster. This fact, which was known neither to Newton nor to Goethe, was one of the reasons for abandoning the corpuscular in favor of the wave theory of light.)

Newton's greatest contribution to optics was his theory of colors, his experimental proof that white light, sunlight, can be separated into, and must therefore consist of, a variety of colors. The device he used was a prism, which splits a white incoming beam into beams of colors identical with those in a rainbow. Surely this must have been the most surprising discovery of the century: Sunlight, the essential unitary agent illuminating the earth, perhaps the greatest (and the first, according to the biblical account) gift of God to mankind, the medium maintaining life, was merely a collection of different colors. This, as we shall see, was hard for Goethe to accept. Did it mean that there were as many different kinds of light particles as there were colors? And how many different colors were there—an infinity?

Many scientific successes followed the prism experiment, the discovery of different measures, different indices, of refraction for different colors. Lenses were known to focus light, but the focus of white light was not a clear white spot; it appeared at different distances from the lens for the different color components. This effect, known as chromatic aberration, is troublesome in connection with the use of lenses, and it confuses the clarity of objects seen through telescopes made of lenses. Recognizing this, Newton conceived a telescope that did not involve refraction but only reflection, which is the same for all colors, thus avoiding the complication of chromatic aberration. As always in normal science, his discovery led to technical improvements, in this instance to the Newtonian telescope.

But there was one ingenious observation that troubled him. When he placed the curved surface of a plano-convex lens (one side of which is plane, the other convex) upon a plate of glass he saw colored rings surrounding the point of contact between lens and glass. There is no way in which this simple arrangement could redistribute the colored particles into circles of different radii, from which they were seen to be reflected. These circles, called Newton's rings, defied the corpuscular theory he had adopted and were later explained by the wave theory of light. But Newton himself found an ingenious explanation for the effect. He endowed the light particles with a most peculiar property: when meeting a surface, they were sometimes disposed to be reflected, sometimes to be transmitted. He sensed that a periodic alternation in the occurrence of reflection and transmission of a given particle was the only mechanism that could simulate interference (which is a property of waves and compatible with today's undulatory theory of light) and thereby produce the effects observed. Thus he ascribed to his particles "fits of easy transmission" and "fits of easy reflection."

What he meant by these fits is difficult to conceive, and the terms as well as the idea were attacked—indeed, ridiculed—a century later by Goethe. But the history of physics has made Newton's seemingly inept conjecture one of the triumphs of his work. In the terms of his time he evidently had the following picture: he seems to have felt that the particles moved in a medium of unknown character and through their motion produced waves in it. A particle at the top of a wave would be disposed to be transmitted, at the bottom of a wave it would be in a "fit of easy reflection." His picture seems to have been that of a row of ducks swimming on a placid pond, ducks that, while swimming, were carried up and down by the waves they had themselves created.

The picture seems fantastic, yet, curiously, it comes close to our present view of the nature of light. It reminds us of the early Copenhagen interpretation of quantum mechanics, when electrons and photons were regarded as both particles and waves. This view, as already mentioned, is now generally abandoned because we have learned that ultimate onta cannot be visualized.

We now turn to Goethe's work. Contrary to the belief of many physicists, it is a major work of three volumes. The first contains 651 pages, of which nearly 300 are a criticism of Newton. The second volume of 757 pages presents an historical, detailed review of the theory of color from Pythagoras to Newton; ten pages of it describe Newton's personality. The third volume contains colored tables and diagrams with appropriate comments.

Before discussing details we state the basic characteristic of Goethe's approach. It is nonmathematical—Goethe scorns mathematics and is generally ignorant of it. He never worries about the nature of light and does not commit himself to Newton's corpuscular or Young's and Fresnel's wave theory. He makes no attempt to use rules of correspondence in order to translate luminous sensations into quantitative constructs. Hence his theory may be called subjective; it confers major emphasis on psychological effects of color, on aspects of interest and use to artists, on features unrelated to technique and application.

It is commonly believed that Goethe misunderstood Newton's prism experiments, that he used a prism borrowed from a friend hastily, expecting to see colors. He did not insert a slit but looked at a white surface, and he found to his amazement that the surface remained white, contrary to what he expected from his understanding of Newton's claim, and that colors appeared only at the edges of the white surface. This convinced him of the unitary, elementary, and

fundamental nature of white light and of the derived, subsidiary role
of colors.

Incidental to his conclusions may be the fact that he began to
write his theory of color after his return from Italy, where, according
to his other writings, he had reveled in the beauty of the colorful
landscape, a beauty conferred by the light of the sun.

The subjective element in Goethe's views is reminiscent of the ancient Greek philosophy which accounts for vision in terms of the
meeting of visual rays, one emitted by the seen object, the other by
the eye. Hence his famous lines, so difficult to translate:

> Wär nicht, das Auge sonnenhaft,
> Wie könnt' es je die Sonn erblicken;
> Lebt nicht in uns des Gottes Kraft
> Wie könnt' uns Göttliches entzücken.

Rough translation:

> If the eye were not akin to the sun,
> How could we ever see the sun;
> If God's spirit did not live within us,
> How could the divine inspire us?

Furthermore, to quote him: "From indifferent, subsidiary animal
features light produces an organ of its own essence; and so the eye
forms itself from light for light, in order that the inner light can meet
the outer."

Fundamental to his thesis is the claim that colors are produced by
the interaction of light and darkness. Sunlight darkened by clouds
gives rise to the colors of the sunset. This he calls the archphenom-
enon (Urphänomen), of which he cites numerous other examples: the
smoke rising from a chimney, the color of transparent material, the
gray appearance of clouds. In a certain sense this belief, the central
idea of Goethe's theory of color, combines all our experiences of
color into a single, orderly hypothesis. To quote Heisenberg, "The
order, which in Goethe's theory of color is constructed before us har-
moniously and filled to its last details with living content, comprises
the entire domain of objective and subjective appearances of color."[1]
In this context Goethe's terminology, his reasoning, are extremely
complex, but his examples are always in accord with known facts.

Subjective colors (Goethe calls them physiological) are an important part of his entire theory. They are properties of the subject, of
the eye, and reveal to us what he calls the chromatic harmony, the

beauty of adjacent complementary colors. He discusses the anatomy of the eye and its reaction, first to black-and-white, then to gray and to colored pictures. Typical is the following suggestion: Look at a piece of blue paper on a white background. The eye sees the blue. Now look away. The eye sees yellow. The eye is provoked to react on its own, producing a contrast that restores a living totality.

At this point he introduces his color circle. Diametrically opposed are those colors "which provoke each other mutually within the eye—yellow-violet, orange-blue, purple-green, and vice versa." Here and elsewhere his scientific discourse is often interrupted by personal reminiscences. His concern with complementary colors is great, and he gives many examples of their production. Considerable space is devoted to "pathological colors," including color blindness.

Although he contradicts Newton, he presents his own theory of diffraction. He distinguishes two cases: (a) refraction without appearance of color. Among the instances he cites is seeing a whole object through a rectangular vessel of water. The object is merely displaced. In case (b) colors appear. Here his explanation is complex but is based on numerous correctly described examples. Refraction with color occurs when white light encounters a bright edge, then a dark plane, or a dark edge and then a bright plane. Newton's prism experiments are ignored here, and Goethe gives an elaborate, today uninteresting theory involving the concepts of primary and secondary pictures.

In his discussion of refraction he does not ignore the prism. He is aware of the fact that different glasses have different powers of refraction and explains the essentials of achromatism produced by a combination of different prisms. But throughout his discourse he continues to insist that only edges produce color. His polemic is extremely detailed and offers a discussion of numerous experiments conducted by Newton. He ends with this summary:

No matter how firmly scientists have hitherto believed to have captured the nature of color, no matter how clearly they imagined to formulate and prove it in terms of a definite [secure] theory, this was by no means the case. On the contrary, they placed hypotheses at the top of their system of reasoning according to which phenomena could be reduced artificially and succeeded in leaving us with a strange [wunderliche] theory of insufficient content.

Another quotation illuminates Goethe's subjective, largely immanent, attitude:

Hitherto light has been regarded as a kind of abstraction, as something existing and acting by itself and, as it were, conditioning itself, an entity which under certain circumstances produces color out of itself.

He deplores the neglect of the eye's intervention.

His polemic against Newton is sometimes severe. He accuses him of being forced "to make an enormous ado about his artefact,[2] to pile experiment upon experiment, fiction upon fiction in order to deceive [*blenden*] when he is unable to convince."

And in one instance it is derogatory and unreasonable, evoking a derisive reaction. Newton says, correctly,

All bodies which are illuminated by complex light (many colors) appear blurred when seen through prisms . . . and display different colors; but those which are illuminated with homogeneous light appeared neither less distinct nor differently colored than they do when directly seen with the naked eye.

Goethe replies: "The eyes must be extremely bad, or the senses entirely fogged in by prejudice if one wishes to see or speak in that way." And finally he gives a general condemnation by stating what he conceives to be the essential difference between his and Newton's approach to an understanding of reality in the domain of color:

Newton's theory has only the semblance of being monadic and unitary. He places his unity at the beginning into a manifold which he wants to extract from the unity, while we develop and construct the manifold out of the postulated duality (of light and darkness).

And in a personal attack on Newton he calls his character inflexible [*starr*], uninterested in what Goethe calls *sittliche* consequences— i.e., personal reactions to color.

All through Goethe's work on color there is a sense of human relevance rather than construction in accordance with what we have called guiding principles for the establishment of physical reality. One section of volume one is entitled *Sinnlich-sittliche Wirkung der Farben.*[3] Here Goethe names each color and describes its psychological effect on the beholder in a most impressive way. Here are four examples.

Yellow: pleasant, alert, softly exciting.

Red-yellow: same as yellow, but more effective. It gives to the eye a feeling of warmth and cheerfulness. It represents the beauty of the sunset.

Green (yellow + blue): our eye finds real satisfaction.

Red: the effect of this color is as unique as its nature. It produces the impression of devotion (*Ernst*) and dignity, of confidence and attractiveness.

Finally, in a chapter entitled "Totality and Harmony," Goethe expresses what is perhaps the principal constituent of his creed in these words:

When the eye perceives a color it is at once activated in accordance with its nature to produce immediately, unconsciously but necessarily, another color which together with the former contains the totality of the chromatic circle. A single color excites, through a specific reaction in the eye, the tendency toward generality.

One of the greatest theoretical physicists of our century has honored the memory of Goethe in an article devoted to the contents of this chapter. Heisenberg[4] is impressed by the poet's intuition, by the way in which the impressions of a sensitive mind almost of themselves combine into a scientific order, how intuition develops immediately from directly experienced nature certain concepts that form the basis of a unique conception of reality. He explicitly states that one does not gain much insight by examining which of the two theories, Newton's and Goethe's, is correct in an ultimate sense. He holds that, while the telescopes and microscopes of today owe their existence to the mathematical theory of Newton, many painters have profited from Goethe's views. And he does not hesitate to call the two theories of color two wholly different layers (domains) of reality. In one of them events and experiences proceed in accordance with firm laws, even when events seem accidental. In the other, Heisenberg says, what happens is not counted but weighed for its human importance, is not explained but interpreted.

He accepts the thesis that reality may be divided into different domains, two of which are the realms of Newton's and of Goethe's analysis. He must, he says, renounce in many places "the living touch with nature" when advancing in the realm of exact natural science.

In the conclusion of his memorable discourse Heisenberg likens the modern scientist, who abandons the realm of live intuition in order to discover the large connections of theoretical discovery, to a mountain climber who wishes to conquer the highest peak of his territory in order to see the general features of the landscape below. He must leave the fertile valleys of his fellow men. As he ascends the mountain, his vision of the land below is enlarged, but the life that surrounds him becomes increasingly paltry, meager, and rare. Finally he reaches a shiny, clear region of ice and snow in which life is non-existent and he himself finds it difficult to breathe and live.

We conclude this chapter by reaffirming the need to retain and appreciate both realities, that of Newton and that of Goethe.

15 The Domains of the Parapsychologist

"And anyone who at the present day expresses confident opinions,
whether positive or negative, on ostensibly paranormal phenomena,
without making himself thoroughly acquainted with the main methods
and results of the careful and long-continued work [of Psychical Re-
search] may be dismissed without further ceremony as a conceited ig-
noramus."
—C. D. Broad[1]

Parapsychology is a field of science of tremendous and
exciting possibilities; a field that numbers among its proponents scientists of such respected eminence and capability in their own fields
as William James, Gardner Murphy, Alistair Hardy, W. Crookes, Ri-
chet, Binet, H. Bergson, W. F. Barrett, G. Murray, and a number of
Nobel laureates. Yet it is largely dismissed by most scientists today as
either a priori impossible and the area of charlatans and dupes by
definition, or else of so little interest that the data are simply not
worth examining and evaluating.

Since approximately two-thirds of all nerve fibers entering the
human central nervous system come from the eyes, and ". . . since
the eye detects primarily patterns of spatial extension, man first perceives his universe as a collection of material objects."[2] Although
human beings have the capacity to see the world in quite other ways,
this is the view that enables us, at least partially, to survive biologic-
ally.

Today we are reinforced in the view that the universe is a collection of material objects by the fact that the last period of our culture
has been what the sociologist and philosopher Pitirim Sorokin[3] has
called a "sensate" period, a time when the cultural view of reality
was that truth is revealed only through the senses. This combination
of biological and cultural mutual reinforcement has made it very difficult for us to believe that the data of parapsychology can be valid.
These data by definition contradict the view of the sensory realm
and, thus, the basic idea of our culture that it is *this* realm that presents truth, so that any contradiction of it is a priori false.

[205]

In 1963 the mathematician Warren Weaver reviewed the work of J. B. Rhine, the founder and foremost experimenter of the modern science of parapsychology, and attested to the soundness of Rhine's statistical procedures and personal integrity. Weaver concludes by writing:

... this is a subject that is so intellectually uncomfortable as to be almost painful. ... I end by concluding that I cannot explain away Professor Rhine's evidence and I cannot accept it.[4]

When asked about the possibility of telepathy, the 19th-century scientist H. L. F. von Helmholtz said: "Neither the testimony of all the Fellows of the Royal Society, nor even the evidence of my own senses, would lead me to believe in the transmission of thought from one person to another independent of the recognized channels of sense."[5] When a scientist of this very high caliber states that he has closed his mind in advance and that no data could make him change it, we begin to see the awesome power our assumptions about the nature of the universe have on us.

The fact that a different model of the world from that used in the sensory realm is called for by the data of parapsychology is implied in an incident reported by Ernest Jones in his biography of Freud. Shocked and startled that Freud considered telepathy to occur Jones exclaimed, "If we are prepared to consider the possibility of mental processes floating in the air, what is to stop us from believing in angels?" Freud replied, "Quite so. And even *der liebe Gott.*"[6] Although the data of parapsychology do not necessarily lead us to a belief in angels and a loving God, Freud was not underestimating the size of the problem.

The conflict created by the data in this field is illustrated by a very large variety of incidents and quotations. To choose one almost at random: When J. F. Coover and L. T. Troland, both outstanding, highly experienced experts in experimental psychology, did a telepathy experiment, they made a mistake in the analysis of their results. Instead of the highly significant results they obtained, they reported the results as due to chance.[7]

So strong have been the psychological and sociological reasons for the rejection of the paranormal in our time that G. N. M. Tyrrell president of the Society for Psychical Research and extremely knowledgeable about the field, could write:

When one considers how far-reaching and important the consequences of this paranormal evidence are, one immediately asks why every inquiring mind has not been filled with a desire to probe them to the limit. Evidence collected in psychical research points to a whole vista of highly significant things—knowledge conveyed without the help of the senses; the future in some inscrutable way open to human knowledge; depths of the subconscious which extend the human personality indefinitely. Why does not even the barest hint of such things excite the interest of every intelligent person and fill him with an urge to explore? The evidence already collected is far more than a bare hint. For three quarters of a century evidence has been carefully collected and reinforced by experiment; yet the reaction to it is not an eager desire to carry the work further and learn all that can be learnt from it. Such interest as there is is casual and the majority tend to explain the evidence away rather than to augment it. There is not the slightest tendency to regard it as a foundation on which to build some of the most important products of human knowledge.[8]

This statement by a leading authority on psychical research does not present an exaggerated picture of the situation; to the contrary, it is more probably an understatement.[9]

The strength of the feelings on the part of the majority of scientists that this field is beyond the pale and must be rejected at all costs has long been known to those who worked in it. In 1930 the psychiatrist Walter Prince documented in detail the fact that when scientists criticized research in the area of parapsychology, they did it in nonscientific ways that they never would have dreamed of using in their own fields.[10] Prince detailed how argument and criticism in their own fields generally followed the conventions of scientific discourse that we have discussed, but when the same people crossed what he called "The Enchanted Boundary" into the area of the paranormal, their arguments and criticism seemed much more ruled by hysteria and emotional feeling than scientific interest.

This problem exists as much today as it did when Prince made his observations. A small incident may illustrate this. In January, 1979, we sent a letter to the journal Science,[11] pointing out that it was generally believed that the type of data studied by parapsychologists contradicted some basic and well-established laws of science, of the stature of the second law of thermodynamics or the law of conservation of mass, energy, and momentum. But we were unable to find any laws of this caliber that were contradicted by the basic findings of parapsychology.

Having heard nothing from Science after six months, we wrote a letter of inquiry. We received no reply, nor did we after sending two follow-up letters (all containing the usual stamped, self-addressed envelope). Finally a personal letter was sent to H. Abelson (then-president of the American Association for the Advancement of Science, which publishes *Science*). This elicited a response from an editor of *Science* saying that our letter had been rejected because we placed the burden of proof upon the critics of parapsychology rather than upon its supporters. This response had absolutely nothing to do with the content of our communication.

The development of the field of parapsychology has been seriously hampered by the widespread use of the concept of reductionism in three areas. The first of these has been the attitude of most scientists that if psi could not be reduced to physical terms—it does not exist. We define a "psi occurrence" as the detected possession of information held by an individual who could not have acquired this information by means of the senses or by the extrapolation of information gained by sensory means. In a careful analysis of this attitude the parapsychologist John Beloff has pointed out that most scientists who reject the possibility of the existence of psi do so because they ". . . see no way of reconciling it with physical theory . . . they assume that what cannot be explained in physical terms does not exist."[12] We need only point out here that if the same rule were followed in psychology, all of Freud's concepts (i.e., 'sublimation') would have to be rejected, for there is no way to reconcile them with neural processes. Nor, for that matter, can consciousness be reconciled with neural processes or otherwise explained in physical terms. However, those scientists (the Behaviorists) who have rejected consciousness on these grounds would be—if they were taken seriously—laughed at by the average intelligent person who *knows* he is conscious.

A second hampering use of reductionism in parapsychology has been pointed out to us by Karlis Osis, Research Officer of the American Society for Psychical Research.[13] This is the general tendency of those working in the field to "explain" all their data by reducing them to extrasensory perception (ESP) and psychokinesis (PK). Whether or not data such as apparitions and poltergeists can be dealt with in these terms is not the question here; rather, it is the apparent belief on the part of many parapsychologists that they can and must be reduced in this way and that they will then be "explained." This has resulted in a widespread feeling of many in this field that some rather dramatic phenomena such as poltergeists and apparitions do not really exist; if they do, they are "nothing but" ESP or PK, and we will be better off in the long run if we ignore them. (The motivation to ignore these must come at least in part from the fact that they make many parapsychologists feel "unscientific." They, too, feel you cannot have a science if things "hop about," and these large-scale occurrences such as poltergeists certainly do seem to "hop about.")

A few years ago one of us was present when Osis, who is a highly trained and experienced parapsychologist, reported to the Board of Trustees of the American Society for Psychical Research. He described in detail how he and another serious parapsychologist (E. Haraldsson) had witnessed under excellent conditions what was seemingly a case of apportation—a phenomenon in which material objects move from one place in space and time to another without apparently crossing the intervening space. The response of the Board to its own Research Officer was overwhelming: "How could you let yourself be gulled like that?"

The third way that the concept of reductionism has worked to hamper progress in parapsychology has been the almost constant effort on the part of a very high percentage of parapsychologists to reduce ESP and psychokinesis to physical models. In spite of the fact that such students of the field as C. D. Broad and John Beloff[14] have shown convincingly that this could not be done, there has been a very strong tendency to keep trying, and a great deal of time and effort has been wasted in this hopeless quest. Some years ago Louisa Rhine, one of the present-day leading authorities in the field, wrote:

The facts of mental ability already discovered in parapsychology no more fit the current idea of a space-time world than such a fact that ships disappear bottom first over the horizon fit the model of a flat earth.[15]

Nevertheless, parapsychologists continue to attempt to reduce the data of their field into the physical model of the sensory reality and "the current idea of a space-time world."

The language parapsychologists use reveals and reinforces the everyday view of reality. Analyzing this language use, the parapsychologist R. Stanford showed that the term "Extrasensory Perception" indicates that psi is

. . . a particular form of sensitivity on the part of the organism . . . [which] naturally puts an emphasis upon ESP as basically an information receiving capacity [and] . . . would seem to imply that in some sense either a specialized receptor or the brain and nervous system must have the capacity to receive and process information.[16]

The term "sensitive," as applied to individuals who repeatedly show evidence that they possess information not acquired through the senses, is consistent with this. This would seem to most parapsychologists to be good common sense. However, it has major implications. For example, it implies a Cartesian Dualism. The brain "gets" information about the *res extensa*. This information is transferred between two separate minds. The information is "out there" to be observed. There is an "in here" (the mind) and an "out there" (the rest of the world). These implications are that the usual view of how-the-world-works is correct and psi data can be made to fit into it, to be reduced to it.

Most scientists, as we noted earlier, appear to be convinced that the findings of parapsychology contradict some of the basic laws of science. We have not found this to be true, although the findings do contradict the Basic Limiting Principles, the basic laws, of the sensory realm. The laws relating observables in the see-touch realm do not permit clairvoyance and telepathy, and the primary route followed by scientists to resolve the apparent paradox has been to declare that there are no such findings. This approach must eventually fail since there *are* such data and it is not permissible in science to decide to ignore data because, for example, the acceptance of them would be uncomfortable or would call for a revision of theories. A second attempt to solve the supposed paradox has been followed by many parapsychologists. This is to try to show that the data *do* fit the model of reality used in the see-touch realm, to find a physicalist model that can be used to make the data fit our common-sense view of reality. One example of this route is the attempt to use the concepts of quantum mechanics, such as that of tachyons flying backward in time, to explain precognition. These attempts at a physicalist model have all proven unsatisfactory; in fact, as a number of writers have definitely shown, they must prove unsatisfactory, for data that violate the Basic Limiting Principles of a metaphysical system cannot be explained within that system. (We shall return to this point shortly.) From our viewpoint the paradox must be solved in quite a different way.

The data in each domain of experience must be taken on their own terms with no preconceptions except that they are consistent; that they do not contradict each other. Impossible events do not occur. Therefore, if a scientist is faced with the fact that an impossible event has occurred—the daily fare of parapsychologists—the paradox must be resolved. This can only be done validly by redefining reality in such a way that what was previously impossible now becomes possible. If the theory must bow to the brute fact, we must be clear as to what is the theory and what is the fact. Our definition of reality, which decides for us what is possible and what is impossible, is the theory. The laboratory experiment in which the paranormal event was demonstrated is the fact.

This understanding of what is the theory and what is the fact is an absolutely critical point in the study of the paranormal—this point has, in the past, received scant attention from parapsychologists and other scientists. We must confront the question, Where do we get our knowledge of what is possible and what is impossible and therefore paranormal? We have ignored the point that a definition of "paranormal" comes from a definition of reality, and that such a definition is a theory, not a fact.

The view that our definition of reality is a fact, that we *know* what reality is and how it works, is a view that would make both science and philosophy tautological, as they are a questioning and exploring of reality. Technology uses common sense; it reflects an acceptance of a particular view of reality and does the best it can with this view to accomplish our ends. Science, as Robert Oppenheimer once put it, uses "uncommon sense." It is a search for new definitions and understandings. Technology takes the locally accepted definition of reality as a fact: science takes it as a theory.

Once the late Niels Bohr and several other theoretical physicists were considering a "wild" theory that one of them had proposed to account for certain peculiarities observed in nuclear physics. The discussion was sharp, and at one point the author of the theory, somewhat shaken asked Bohr: "Do you think this is crazy?" Bohr pondered briefly. "Yes," he said, "It is crazy, but I think it is not crazy enough!"[17]

The kind of uncommon sense, of daring and questioning of basic definitions needed in science, the kind needed in parapsychology, is demonstrated by a remark of the great mathematician David Hilbert. At one time he praised a new student of his who seemed to show great promise. Some time later Ernst Cassirer asked him what had happened to this student. Hilbert replied, "Oh, he did not have enough imagination to be a mathematician, so he became a poet!"

From this viewpoint the 19th-century philosopher David Hume was in error in his famous argument on disbelief in miracles, and similarly the countless arguments against parapsychology stemming from it are in error. Hume defined his interpretation of how-the-world-works as a fact when it was a theory. As a *fact,* and given the faith of philosophy and science in the consistency of reality, it was blatantly impossible for it to be contradicted by another fact (the paranormal occurrence), and therefore the paranormal occurrence logically never happened, and its observers were mistaken or lying. The chain of logic is unassailable so long as the definition remains unquestioned. Once the definition is examined, however, it becomes clear that it is a theory, not a fact, and that, therefore, when opposed by a fact, it must be given up as inaccurate or incomplete.

We can see the problem clearly when we think about the colleagues of Galileo who refused to look through the telescope. They refused because it was unnecessary to look; they had confused their theory about reality with facts. As far as they were concerned, they knew the facts, and there was simply no point in observing a contradictory fact; the telescope's view was necessarily false because it contradicted known facts. At this distance we can see their reasoning and their confusion clearly. It is, however, harder to see when many modern scientists, not looking at the facts of parapsychology, simply dismiss them as necessarily false and therefore unnecessary to examine since—for them—they contradict a known fact. They are as confused as were Galileo's contemporaries, but this is a lot harder to see close up.

As we have indicated earlier (chapters 2 and 12), the social scientist and the physicist approach the necessity of different metaphysical systems to make data lawful in different ways. The physicist asks which domain of experience we are dealing with, what the observables are, and what laws are needed to make the relationships between the data lawful. The social scientist asks in what way the person being studied is organizing and construing reality at a particular time and what are the laws of that organization. There is no contradiction between these two approaches, but they are quite different. Let us now examine the problem of "paranormal" from the viewpoint of the social scientist.

A theory about reality, a conception of how-the-world-works, which is so real to us that we perceive and react as if it were true, as if it were a fact, can be described in two ways. From one viewpoint it is a state of awareness, a state of consciousness, a way of being-in-the-world. From this viewpoint, the one we have when we are *using* the theory personally, we are responding to the truth about reality. This is how things and we are. From the other viewpoint, it is simply an integrated set of hypotheses concerning reality and is judged by its effectiveness in attaining whatever goals seem relevant to whomever is doing the judging. It is a theory of metaphysics to be compared with other theories of the same kind.

These two descriptions—a state of consciousness and a metaphysical theory—are opposite sides of the same coin. When using them, we are talking about the same thing from two different angles; they are the same phenomena experienced in two different ways. This implies that there is no such thing as a generally "correct" or "normal" state of consciousness; rather, there are various states that can be compared in the way they succeed in aiding us, permitting us, to solve our problems, to arrive at our goals.

We therefore no longer ask which construction of the cosmos, which state of consciousness is the correct one in that, when using it, we are perceiving and reacting to reality. We only ask which construction and which state of consciousness is most effective in helping us attain which goals. The concept of a "correct" or "normal" state of consciousness is one we will have to put on the crowded and dusty shelf marked "Outmoded ideas: Ingest at your own risk." We can, however, ask, "Which state of consciousness is most useful to solve certain needs and goals?" and "Which state of consciousness is statistically most prevalent in which cultural situations?"

This comprehension is—we've tried to indicate in this book—one of the most staggering and least understood insights of modern science. We no longer search for what reality *is*, but rather for ways of usefully construing it; ways to define it that will help us achieve our goals. It is that there is no "right" metaphysical system, but only a number of compatible ones of limited usefulness. There is no "correct" state of consciousness that will reflect "reality," but only a number of states useful or useless for specific human purposes.

The next step follows naturally. If there are a number of different, equally "right" metaphysical systems—states of consciousness—and if these are quite different in the entities and laws (the observables and laws relating them) they contain, we can do certain things with some of them that we cannot do with others. What is "normal" in one of them is "paranormal" in another. For something to be "paranormal" in a particular construction of reality means that it is forbidden by the Basic Limiting Principles of that construction, and so it does not happen when we are using it. It cannot be "explained" in that metaphysical theory since it does not happen in it. One cannot explain impossible events within the metaphysical system (theory about reality) in which they are impossible.

This is central to the problem that parapsychologists have had in "explaining" or "understanding" psi phenomena: If a system of real-ity-ordering forbids certain events from occurring (as, in the sensory realm system, an effect preceding its cause in time could not occur), you cannot explain that event within the system. It is like trying to explain parallel lines meeting within the system of Euclidean geome-try; you can try all you want to do it, but you simply can't. If the event occurs (as in laboratory demonstrations of precognition), you simply have to explain it within a system in which it *can* occur. You can explain parallel lines meeting within the system of Riemannian geometry; you cannot in the Euclidean system. It is not that it is dif-ficult to explain or complex to explain, it is that it cannot be done.

There is an old story about the lost traveler who asked the coun-tryman how to get to Salisbury. The farmer replied, "You go north five miles and then turn west . . . no, that's no good. You go west three miles and take the first road north . . . no, that won't do it. You go east and then . . . by God, you can't get there from here!" Para-psychologists have tried and tried to get from here to there on the solid-appearing roads of our ordinary theory about reality, the theory of the sensory realm. It can't be done. In this realm we can do certain things and we can't do others. We can travel to Yankee Stadium, Waterloo Station, or the Place de l'Étoile. We can't travel to the day before yesterday or to the Land of Oz. You can perceive something with your senses or extrapolate from known data. You can't be clair-voyant or precognitive. You cannot explain events forbidden by a system within that system.

Psi occurrences are not possible in a world that uses clock time and yardstick space. In the metaphysical system that uses these defi-nitions, the metaphysical system needed to make the data of the sen-sory realm lawful, psi occurrences are truly paranormal. They are impossible and therefore do not happen.

However, when we realize that psi events are events in conscious-ness and that consciousness does not use clock time and yardstick space, the paradox begins to be resolved. Consciousness needs per-sonal space and personal time to make its data lawful. In the meta-physical system in which these are facts, psi occurrences are not forbidden; they are not impossible.

There is, in many scientific and pseudo-scientific circles today, a good deal of talk about a "paradigm shift." The idea behind this is that science, and our culture generally, is in the process of changing the model of the universe, "how-things-are-and-work," from the me-chanical world-picture to a new one. The trouble with this talk is that it rests on the old idea that there is just one "paradigm," which includes the entire cosmos. Since Quantum Mechanics and Relativity Theory, however, this is no longer valid. We do *not* need a new model of reality in the sensory realm, for example. The old one is perfectly adequate. We did need new ones in the microcosm and in the macrocosm. We developed them. We do need new ones in the realms of consciousness and parapsychology. We can develop them. These models will be compatible with the others in use in science, though they will be different. We do not need a "new paradigm" to cover the cosmos. We need rather to look at each domain of experience and to see what description of reality fits the data.

From the viewpoint of this comprehension the spiritualists and theologians were more correct than the scientists when they tried to explain paranormal events by saying that spirits produced them or that God produced them. They were taking entities from another metaphysical system to explain phenomena inexplicable in the sensory one. They were thereby implying that what was needed for the explanation of paranormal events was a different metaphysical system, a different state of consciousness, while most scientists and parapsychologists tried to hold onto the see-touch metaphysical system and explain them in it, where the events were forbidden and therefore their explanations forbidden.

We say that the spiritualists were "more correct," but not that they were "correct." The situation is similar to that of the little boy who came home and told his mother he had gotten first prize in an examination. The question asked had been, "How many legs has a horse?" He had answered "Three." When his mother asked how he had gotten the first prize, he replied that all the other children had said "Two."

Let us turn now from the approach of social scientists to the general approach physicists use when dealing with major problems. Physicists ask what is the domain of experience in which they are working, what are the observables found in this domain, what laws are necessary to make the relationships between these observables lawful. We will deal in this essay with one set of observables found in parapsychology; what is generally called in the field "psi-gamma" occurrences, or ESP. These are cases of the detected possession of information by an individual who could not have acquired the information by means of the senses or by manipulation of information acquired through the senses. (We will not here discuss the problem of "psi-kappa" occurrences or psychokinesis, but will leave this for later work in other places.)

We do not intend here to review the data of parapsychology. This has been done admirably elsewhere in a large number of publications. Both authors have satisfied themselves that the data are real and not due to artifact of research design or to chicanery.[18] From a scientific viewpoint the problem is no longer one of the existence of psi-gamma events, but one of how to develop a theoretical model by means of which to relate them to the general framework of science.

The possibility of a scientific methodology for the study of the type of events that have been historically the major interest of psychical research[19] does exist. Although an excellent and productive methodology has been devised for one type of psi occurrence (the type studied in the laboratory by card-guessing experiments and the like), the dramatic events that originally stimulated interest in this field have gone without one. In this area psychical research has remained largely in the anecdotal and descriptive stage of development.

As an example of these large-scale paranormal incidents, let us consider the following case:

In 1930 a one-eyed pilot named Hinchliffe attempted the first east-to-west transatlantic flight. He had expected to fly alone. Unexpectedly, at the last moment, his financial sponsor insisted on a woman copilot. Several hundred miles away, on an ocean liner, unaware that Hinchliffe was making the attempt at this time or that there were any plans for anyone to be with him, two old friends of his, Air Force Colonel Henderson and Squadron Leader Rivers Oldmeadow, are sleeping. Then, in the middle of the night, Henderson, in his pajamas, opens the door of Oldmeadow's cabin and says:

"God, Rivers, something ghastly has just happened. Hinch has just been in my cabin. Eye-patch and all. It was ghastly. He kept repeating over and over again, 'Hendy, what am I going to do? What am I going to do? I've got the woman with me and I'm lost. I'm lost.' Then he disappeared in front of my eyes. Just disappeared."

It was during this night that Hinchliffe's plane crashed, and he and the woman were killed.[20]

This is the type of data that historically have been the primary concern of psychical research. The information that Henderson reported was meaningful and important. Unfortunately, very little progress has been made in the past hundred years in increasing our understanding of this type of phenomenon.

The psychiatrist and parapsychologist Ehrenwald, in an important paper,[21] differentiated two types of psi occurrence. The first referred to "psychologically significant and dynamically meaningful incidents ... of a purposeful, goal-oriented nature." They are *need-determined.* (There is a *need* to communicate, and "ordinary" sensory communication is blocked.) The information in the experience is perceived as important and significant to the "receiver."

The second type occurs without conscious awareness. It is "structurally rather than dynamically determined ... [and] due to a cluster of neurons caught napping at their jobs or to the irregular firing of others." It is *flaw-determined,* due to a temporary and local breakdown of the "filter" system that keeps us from being overwhelmed with psi-transmitted information. It is "facilitated by such minus functions of the ego as REM sleep, relaxation, sensory deprivation ...." The information is erratic and not perceived by the individual consciously, or if its existence is pointed out, the content is seen as unimportant. It is not psychologically significant to the individual.

These two types appear to be overlapping areas of a spectrum rather than clearly and sharply separate types of psi.

We are concerned here with Ehrenwald's "significant," "need-determined" type of psi occurrence, of which the Hinchliffe-Henderson story is an example. It is this type of occurrence that has classically been the central interest of psychical research. However, in this field's major attempt to become "scientific"—the shift in emphasis that we might term the change from "psychical research" to "parapsychology"—many of the research workers involved felt that it was necessary largely to abandon this type of occurrence as objects of study and to concentrate on Ehrenwald's "flaw-determined," "leakage" type of occurrence. Typical of this concentration was the tremendous number of statistical studies of card-guessing experiments. A good deal of progress has been made in recent years in studying this second type of phenomenon.[22] Very little, if any, however, has been made in the study of the "need-determined" type.[23]

To begin the approach we are suggesting here, let us decide on the "domain" with which most psychical research is concerned. This is the cross-section of experience in which there is more than one human being. Simply put, it is in this domain that we *observe* psi occurrences. Further, although it appears possible to conceive of pure clairvoyance or precognition in a one-person cosmos, this calls for a great deal of intellectual stretching. More important, however, is the fact that it is almost impossible to conceive of a person, a human being, developing or existing as such, alone in the universe. If, as the psychologist W. Köhler once wrote, "A solitary chimpanzee is not a chimpanzee," how much more obviously true is this of a solitary human being? A voluminous literature in psychology and psychiatry bears clear witness that human psychological characteristics only develop in the working out of relationships with other persons.[24]

In this domain of multiple human beings we find three classes of observables "appearing": They are "self-aware individual identity," "communication," and "relationships between people." There may well be other observables we will wish to include later, but these will suffice for us at this time. If we wish to follow the classical model of the successful sciences, our primary question will be: How do these observables relate to one another? We will, along the course of this way, find ways to define our terms carefully. However, for the purpose of demonstrating the possibility of this scientific model for psychical research, we can, at present, be content with doing this in a general and rather loose way. "Communication" we will define for now as the detectable transmission of information between two individuals. We shall divide this into two kinds: sensory communication, in which the transmission takes place through the sensory organs or through manipulation of information that was acquired through them; and nonsensory communication, or psi occurrences. We are (following the parapsychologist Charles Honorton) inserting a "detectable" into both of these definitions, since nondetectable entities or processes are of no interest to science. (It may well be, for example, that psi transmission of information always or usually accompanies sensory transmission of the identical information. If true, however, this would not be detectable, and science takes it as a general operating rule that entities that are in principle nondetectable are to be treated as if they do not exist. See, for example, the history of the concept of the "ether.")

Psi occurrences are only detectable when sensory communication between those involved is blocked. (Otherwise, whether or not psi is occurring, one attributes the communication to the sensory interaction.) For our purposes, then, psi occurs when sensory communication is blocked. Since our interest here is in the "need-determined" type of occurrence, there obviously must also be a need to communicate on the part of at least one of the persons involved.

Let us now begin to make hypotheses as to what the connections are between these three observables—communication, relationships, and identity. So far as communication is concerned, we are in-

terested, as already indicated, in the type that occurs when information transmission through sensory systems is blocked and there is a
need to communicate. Such communications—Ehrenwald's "need-
determined" type—are of events important at least to one of the individuals concerned.

Let us look first at the observable "relationship." Do we already
know anything about this that can be of help in formulating testable
hypotheses? It turns out there is a good deal that we already know.
From the research by psychologists into small-group behavior—for
example, from the Group Dynamics of Kurt Lewin and his students
and from the Interaction Process Analysis of R. F. Bayles and his followers—we can make some definite statements. (It should be borne
in mind that the "small group" starts with and includes the dyad—
two people relating to each other.)

There is, for example, a measurable attribute in relationships generally called "cohesion." This has been defined as "the total field of
forces which acts on members to remain in the group[25]—or, in a
dyad, to continue the relationship. (An observable in science may
have attributes as the observable "force," in physics, has the attributes "strength," as measured in number of dynes, and "direction."
Cohesion is analogous to the attribute "strength of force" in physics.)

Using cohesion as the primary variable, the following conditions
(secondary observables) are among those that have been shown to affect it:

A. Cohesion is greater when the emphasis in the group has been
   on cooperation rather than competition.
B. Cohesion is greater in a democratically organized group than it
   is in a group governed by authoritarian or laissez-faire procedures.

Our first hypothesis then might be that *psi occurrences are more
frequent between individuals whose relationships have been coopera-
tive than they are between individuals whose relationships have been
competitive.* Our second hypothesis might be that *psi occurrences are
more frequent in egalitarian than in authoritarian groups.* (Although
testing these hypotheses would be difficult and require various "cor-
rection factors" for bias, the tests themselves are perfectly feasible.)

Since the stronger the interpersonal attractions among its members, the greater the group cohesion,[26] we can make a third hypothesis. *Psi occurrences are more frequent between people who like each
other than between people who do not.* (It is of interest that a recent

study by Carl Sargent has shown that parapsychologists who get good experimental results are more likely to be open, warm, and friendly than those who do not.[27]

Our reasoning in developing the above hypothesis is not complex: The greater the cohesion of a relationship or a group, the greater will be the tendency to continue communication in the face of difficulty. The blockage of sensory communication is one such difficulty. Psi occurrences of the type we are interested in here are a way of continuing communication when sensory channels are blocked. Therefore, whatever increases group cohesion will increase the predicted frequency of the occurrences. There are known to be other testable variables that affect cohesion as well as measures of cohesion itself. We could thus make other predictions of the above kind, but these three examples, should adequately demonstrate this possibility for present purposes.

Research on the observable "communication" has given us data indicating that important interpersonal transmissions of information (transmissions that change the persons involved in a way perceived by at least one of them as significant) tend to take place between individuals identifying themselves as members of the same group.[28] We would therefore make the following hypothesis: *People who identify themselves as members of the same, important (to them) group, will report psi occurrences between them more frequently than those who do not.*

Social class in the United States is a system that tends to separate individuals into different groups, life-styles, and patterns, and an individual's position in the social-class structure is generally determinable. We might therefore make this hypothesis: *Psi occurrences between two members of different social classes will be reported much less frequently than between members of the same social class. (This, however, will not be true if there exists a special group that includes both of them and is important to at least one of them.)* The implication of these two hypotheses for the psychological design of psi experiments and laboratory behavior of staff is obvious.

The social psychologist R. F. Bayles and others of his school have approached communication primarily from the viewpoint of problem-solving activity. They have demonstrated, for example, in a large number of experiments that human beings need and strive for stability (another attribute of the observable "relationship") in their dealings with others and develop "roles" to maintain this stability. Solutions to problems of interaction become institutionalized as roles so that stability (and therefore predictability) can exist. (There are, of course, other reasons for the development of roles.) Bayles has demonstrated the consistency and importance of this aspect of relationships.

A "role" can be approached from both a sociological aspect ("He is a father to those children") and from a psychological aspect ("He is a very demanding father"). There is a strong tendency for roles in a group (including a dyad) to be consistent and communications to be relevant to them.[29] We might therefore make the following hypothesis: *A psi occurrence will be in keeping both sociologically and psychologically with the role that the "agent" plays or has played in relation to the "percipient."* A second hypothesis might be that *psi occurrences are more likely when the stability of an important relationship is threatened and communication is necessary to maintain it, but sensory modes of communication are blocked.*

In his work Bayles has developed a method of analyzing verbal communications in a relationship into four general classes: positive reactions; attempted answers; questions; and negative reactions. The first three classes indicate that the predominant forces operating at this time are those favoring the continuance of the relationship, the fourth favoring its discontinuance. In terms of our earlier comments about cohesion we might make the hypothesis that *verbal communications preceding psi occurrences are more frequent in the first three classes than in the fourth.*

These hypotheses are concerned with the mutual variations between the observables' "relationship" (specifically its attributes cohesion and stability) and "communication." The same type of hypotheses can be made concerning the variation between psi occurrences (the type of "communication" we are interested in here) and "identity."[30]

The interactions between relationship and identity have been widely explored in a large number of contexts within both scientific and artistic and literary frameworks. It has long been clear that one cannot exist without the other, and it does not seem necessary here to describe the large body of literature that is unanimous on this point.[31] Further, it has become clear that although there may be shorter or longer periods in the individual's life when direct communication is cut off (the Robinson Crusoe situation, for example), the three observables of identity, communication, and relationship are as interdependent as are volume, pressure, and temperature in another domain.[32] It is the consciousness of a relationship or membership in a group that is important in determining identity and behavior. "Groups have a consciousness of membership which may, indeed, persist even when intercourse with co-members has ceased, as with an Englishman living abroad."[33]

Let us here use as one aspect of the observable "identity" Erik Erikson's definition that identity is the ability to maintain important patterns in the face of change.[34]

Since the evidence from the literature is clear that people strive to maintain their identity with the same intensity and need as they do to maintain their relationships, we can make certain hypotheses about the interaction of "identity" and the frequency of psi occurrences. One example of this would be as follows: *The psi occurrence will tend to aid the individual in maintaining important patterns in the face of change.* Restated, this hypothesis would read: *"Psi occurrences will tend to stabilize identity and maintain consistency of action and perception more often than they will tend to destabilize identity."*

All the hypotheses we have presented here seem to have much in common. This is because we are dealing with a Gestalt[35] of identity, communication, and relationships, and separating them is artificial and only for the purposes of making it possible to make the hypotheses testable. Ernst Cassirer has pointed out, for example, that a major function of language is to ensure that a group has a common experience of reality and that the participants are enabled to communicate, relate, and maintain their identities.[36]

Those sciences that have been able to make definite progress in the past period followed courses containing similar steps and procedures. These included *the selection of a specific domain, the identification and definition of the observables in this domain, a concentration on the question of the relationships between these observables.* In the history of psychical research it was not believed possible to do this with certain types of data. These types included the meaningful, "need-determined" psi occurrences.[37]

Scientific procedures in the field, therefore, focused primarily on the less personally meaningful type of psi occurrence that Ehrenwald has labeled 'flaw-determined.' Methods were developed by J. B. Rhine and his followers to study these with careful and precise scientific procedures. We might loosely characterize the change from "psychical research" to "parapsychology" as being a change in emphasis from the study of mediums and of death-bed apparitions to the study of statistical guessing of randomized packs of cards.

Our purpose here has been to demonstrate how domain theory is relevant to, and can be applied to, the field of parapsychology. In doing this we were able to show that a scientific approach could be made to the area of need-determined psi occurrences and to give some basic examples of the kind of hypotheses that could be made and tested. The presently existing rich literature on relationships and on identity as well as the extensive literature on psychologically significant psi occurrences gives us a wide field for making and testing other hypotheses in this domain.

16 The Domain of Ethics

"THY WILL BE DONE on earth as it is in heaven."

This citation from the Lord's Prayer implies that man should obey God's commandments. They are usually taken to be His ethical commandments, which we shall in the context of this chapter call laws of ethics. The sense in which they are laws will be made clear as we proceed. They are not as simple and as universal as the laws of science, having different formulations in different cultures and religions, but they act as laws nonetheless.

There is, however, this affinity between the ethical laws and the scientific laws. The above quotation suggests that God's will created the ethical laws, whatever they are conceived to be. The parallelism between their origin and that of the laws of nature is not generally understood. To set matters straight, at least for the Judeo-Christian mind, we present here another quote from the Bible, which shows God to be responsible for the laws of nature as well.

The usual story of the six-day creation of the universe is emphasized in all biblical teaching, while the story of God's affirmation of lawfulness, which occurred at the end of the great flood, is rarely interpreted as a promise of consistency and regularity in his conduct of the world, as the creation of the laws of nature. Recall the promise Jehovah made to Noah when the rainbow appeared over the waning flood, which terminated the turbulent days, the *tohu vabohu* of the preceding era:

Jehovah smelled the sweet savor; and Jehovah said in his heart, "I will not again curse the ground any more for man's sake, for that the imagination of man's heart is evil from his youth; neither will I again smite any more

[224]

everything living, as I have done. While the earth remaineth, seedtime and harvest, and cold and heat, and summer and winter, and day and night shall not cease." And God said, "This is the token of the covenant which I make between me and you . . . for perpetual generations: I do set my bow in the cloud. And it shall be for a token of a covenant between me and the earth. And it shall come to pass, when I bring a cloud over the earth, that the bow shall be seen in the cloud: And I will remember my covenant, which is between me and you and every living creature of all flesh.

The purpose of the foregoing introduction is to remind the reader of the common origin which some religions ascribe to the commandments of ethics and natural laws. In this chapter we present philosophical, indeed epistemological arguments which bear upon this point, arguments that show a degree of similarity in the structures of science and of ethics which might bring them closer together and, we hope, remove the uncertainties and conflicts in our understanding of the principles and workings of ethical systems.

We have seen in this book that consciousness is peculiar, unique, and beyond the traditional scientific approach used to establish physical reality. It lacks rules of correspondence that render its observables objective and quantitative. Ethics is another concern necessary to human existence that lacks scientific rules of correspondence. It comes close to science in its methodology, though it lacks operational definitions, replacing them by equivalent relations in its own domain. As we shall see, it leads to something that might be called "ethical reality," though this term is hardly ever used (the term "validity" comes close to it).

First, there exists a certain parallelism between the structures of normal science and of ethics. We encounter a kind of transcendence with compatibility between science and ethics: the laws are similar and compatible, but the *observables* are different. In one domain these are the ones we treated earlier in connection with the physical sciences as well as in biology (which we have ignored); to mention a few—position, temperature, valence, chemical bond strength, gene structure, brain size, blood pressure. In ethics the observables are *values* with a great variety of names, and they stand in a relation to the former observables which we called transcendence. To see the parallelism we return briefly to the diagrams in earlier chapters and to the realms they signify. Our simplest example was the perception of a falling stone as represented by the sequence of protocol experiences that are indicated in Figure 8 by P_1. For simplicity we draw only a single P_1, although to state matters completely, we ought to

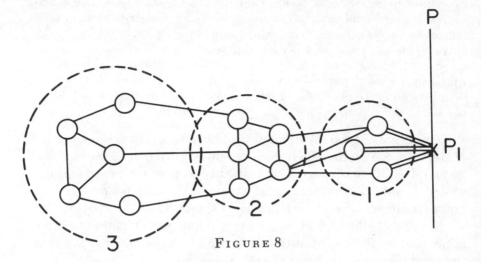

FIGURE 8

draw many. Rules of correspondence (R) lead from P_1 to observables like position, speed, acceleration, and mass. They are related by Galileo's equation for free fall: distance of fall equals one half the acceleration of gravity times the square of the time of fall (provided the initial velocity is zero). This represents a primitive law, a combination of constructs close to the P-plane. Its realm is indicated in Figure 8 by the contour labeled 1.

This relation, however, can be derived from a more general one—namely, Newton's law of universal gravitation—which is more extensible—that is, it relates to a larger realm of P-facts (not drawn in the diagram) than the former. (contour 2) To put the statement in reverse, Newton's law logically implies Galileo's. For several centuries the former was ultimate, could not be deduced from anything more fundamental, and therefore lay on the left boundary of the C-field. But Einstein showed that Newton's law, slightly modified and corrected, was a consequence of his theory of general relativity. In terms of our diagram Einstein established a postulate or an axiom (a logical relation between more fundamental observables including the parameters of the metric of space) to the left of Newton's realm, contour 3. There are of course, many connections between the constructs in 3 and those in 2, as well as rules of correspondence between some of the constructs (circles) in 3 and the P-plane (i.e., single and double lines), which are not drawn.

To say that a given phenomenon is explained means that we can pass continuously from some P-experiences to the left, through the

C-field, until we come to the set of laws called axioms. These differ from those to the right of them because they are the last instances of appeal, not derivable from any more basic propositions (which would lie to the left of them.) Figure 8 represents part of the domain of mechanics.

A similar account could be given for every conventional science, such as electromagnetism, quantum mechanics, atomic and molecular structure, genetics. Each has on the left of its C-field a postulate or a set of postulates beyond which a passage to the left is impossible. They are never absolute: New theories often push to the left beyond what was previously called a basic axiom. This is the essential meaning of the claim that science does not harbor absolute truth.

The question arises: Whence came the postulates? Inductive empiricism holds that they are somehow derivable from P-facts, but most creative scientists today deny that claim. Einstein's conjecture did not begin with a careful analysis of all the "facts": It was a creative act different from a guess only in its success and the aesthetic aspects that mark it a priori as a conjecture of genius. The word "inspired" describes it well.

We now trace in Figure 9 the movement of science from the axioms (A) to various protocol facts (P) through very general laws or equations close to A, then to less extensible laws, and finally to specific relations among observables near the P-plane. If these relations, translated by rules of correspondence (R) into P-facts, are *verified*, we have conducted a successful course of explanation. The intermediate region of laws contains what we often call theories and is labeled T in Figure 9.

We have just asked about the origin of the postulates and found them to be human inventions, but what about the rules of correspondence (Rs)? We have seen in previous sections that they too involve features subject to human choice. We recall that among numerous ways of measuring time, we select the one that leads to Newton's first law of motion; among all possibilities of defining temperature we select the one that leads to the ideal gas law and other simple thermodynamic relations—in the strictest sense, therefore, *verification* in science, the establishment of agreement at the P-plane, involves the postulation of certain rules (of correspondence). Science starts and ends with deliberate human choices. That the scheme, i.e., the particular coherent definition of physical reality works, is in a sense a miracle.

The structure of ethics is formally similar; its language, of course,

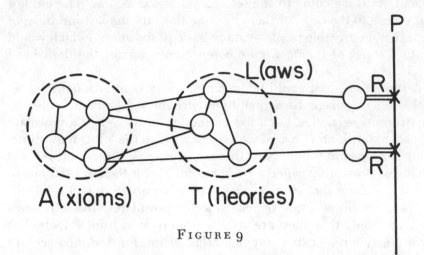

FIGURE 9

is different. While science, describing facts, is largely tied to the indicative mode of speech, ethics, intended to prescribe and proscribe human actions, must use imperatives somewhere in its range. Its methodological structure is exhibited in Figure 10, about which the following must be said.

It has been shown far more explicitly than is appropriate in the present context that every ethical system which has proved viable for a substantial period in history (say about a millennium) has imperatives, progressively amplified and refined into laws, at its beginning, both in time and in its methodological structure.[1] These imperatives—for example, the Ten Commandments, the Golden Rule, and the Eightfold Way—are most frequently regarded as divinely inspired. Looked at in the human perspective, they are postulates.

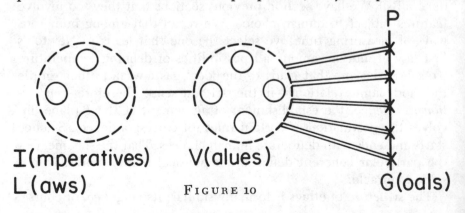

FIGURE 10

Hence at the left of Figure 10 we find commandments, imperatives (*I*), or, in their further explication, codes of law.

Action by a given group of people[2] comprising a tribe, a nation, or culture will result in specific forms of behavior generating constructs usually called "values" and labeled *V* in Figure 10. Thus, the imperative "Thou shalt not kill" makes life a value; a commandment that forbids stealing establishes property as a value; "Honor thy father and mother" establishes the value called filial piety; "Forgive your neighbor's trespasses" creates the value tolerance or charity, and so on. We shall call these values, which arise automatically from obedience to the imperatives, *factual* values in order to distinguish them from others which will be introduced in a moment.

How are the values related to the imperatives? In Figure 9 the relation between axiom (*A*) and theory (*T*) is one of logical entailment, that between *I* and *V* in Figure 10 is established by the process of living in accordance with *I*.

We have called the *V*'s factual values; they arise automatically from the imperatives and have no inherent validity, no compelling force except conformity with the imperatives, which, as we have seen, are largely arbitrary. If in science we began our reasoning with an arbitrary set of *A*'s, we should also arrive at theories by analytic implication, but we could not be sure that they were "true"—i.e., that they describe a useful and coherent picture of reality. Correspondence with protocol facts is necessary for that.

The adjective "factual," which describes the values, expresses a similar arbitrariness. Considering them alone we cannot tell whether their ancestors, the imperatives, were "valid." The *V*'s describe peoples' behavior; they are what the statisticians discover by studying what people actually, and most frequently, do. Too often, the general practice is taken to be a norm. Actions are supposed to be right, or good, if everybody (or the majority) performs them. This fallacy is commonplace and vicious. (It even affects the enforcement of laws.)

The point is that factual values lack normative force; being merely factual, they lack the "ought." We now consider under what conditions the "is" merits and thus acquires the authority of an "ought," how ethics attains "validity," which is the counterpart of "reality" in science.

Curiously, attainment of validity also involves the establishment of correspondences, this time with another set of values, called ideal. Every system of ethics contains, aside from laws and their attendant behavior V, a set of goals, G in Figure 10, which may be called *ideal* values. Strangely, they are almost universally accepted in all cultures; they contain such maxims as human happiness, personal and collective, freedom of action or even belief, life, health, stoic tranquility or the peace that passes understanding, and perhaps a measure of privacy and perhaps marital faith or even nirvana. They include what our politicians call human rights. Despite the variety of names all of them are compatible, even though postulated and therefore to an extent arbitrary. The near-universality suggests divine inspiration just as does the striking similarity of the imperatives. Correspondence between factual and ideal values, the attainment of happiness, freedom, etc., by living in accordance with the commandments is taken to "validate" the ethical system and thereby transforms the actual values into ethical values, the factual into the normative.

Are there any suggestions in this exposition that are useful for the study of consciousness? We have seen that validity and completeness require correspondences (R) in both science and ethics. But while those of science, such as operational definitions, introduce quantitative measures, those of ethics do not—or at least not quite. For one could count the number of times a given action performed by a specified group of people has realized a given ideal and assign that number to the action. Thus all—or at least many—V's would be measurable. Perhaps this method of quantification contains hints for the study of consciousness.

The theory of ethics here outlined, when viewed from the standpoint of a person in a modern Western society and applied to its functioning, exhibits major difficulties. For it is idealized in that it applies to a uniform, unstratified group whose concerns are essentially ethical. Class structure, stratification by personal property, the caste system make it relevant in principle only to a single class, stratum, or caste. It would work in an ideal utopian communist society as it did in early Christian communities and still does in certain American groups held together by religious ties and stabilized through respect for their history.

Our Western civilization is hardly the place to test the validity of any logically structured ideal ethical theory. For it is highly diversified with respect to personal property, which is one among many other values in our scheme, but which confers inordinate power and esteem above all others. It mixes and confuses ethical with economic considerations so that bureaucratic committees on ethics are needed to disentangle them. Its polite name is capitalism; its driving force is called personal initiative, and it has turned into greed, tempered by public, official charity to forestall explosive revolutions. One of the abhorrent examples of the indiscriminacy of ethical and economic political procedures is the willingness of publishers to spend millions of dollars for the rights to books written by criminals.[3]

While it seems useful to point this out by way of being realistic, the essential structure of the ethical system retains its importance, if not as a description of our society, yet as a logically and essentially valid scheme whose methodology contains hints as to the manner in which a subject that cannot be quantified in the manner of traditional science can nevertheless proceed and possibly succeed. What it shows is that the rules of correspondence leading to our understanding of the external world can be replaced by correspondences between what we call factual and ideal values. Whether this observation contains hints for the study of consciousness remains to be seen.

The method suggested in this book for dealing with novel realms of experience involves first of all the selection of meaningful observables, then relating them by laws that explain or predict what we have called P-experiences. To apply it to ethics, we could take P-experiences to be simply *satisfaction of the postulated*—perhaps we could even say universally accepted—*primary values* (the goals in Figure 10, which we shall now call P). Observables are clearly the factual values, V, which originate from behavior in accord with L, the laws. These cannot be measured (quantified) in the ordinary sense. Names for them might be wealth, health, observance of daily duties, diligence, honesty, family harmony, longevity—and their opposites. Connection between them—i.e., laws relating them, or some of them—can surely be discovered. To assign numerical values to them directly is impossible. But for a given society living under given laws the prevalence of each observable can doubtless be determined statistically on a scale, say, from -1 to $+1$.

Notice that the observables have been chosen so that their fullest measure ($+1$ for each) would satisfy the goals, P. If the sum of their measures equals its possible maximum within a reasonable error, the ethical system is successful; the laws are *validated* just as in science the axioms are *verified*, also necessarily with tolerable errors.

We complete this chapter by restating an allusion that pervades it and is summarized in Note 3. It appears that our democracy is perhaps biased in so far as it confers greater emphasis on human rights than on human duties. This view is dramatically expressed by the distinguished logotherapist V. F. Frankl at the conclusion of his popular book: "I recommend that the Statue of Liberty on the East Coast be supplemented by a Statue of Responsibility on the West Coast."[4]

17 The Domain of Consciousness

THERE IS A VERY OLD IDEA in Western knowledge: *Natura non facit saltus*—Nature does not make leaps. This idea, that all nature is continuous and that all differences shade into each other in a series of infinitely fine gradations, is one of the premises giving rise to the error of reductionism, which, as we've seen earlier, insists that honor is "nothing but" a reflex we are conditioned to by our cultural upbringing, that love is "nothing but" the biological urge to reproduction, and that consciousness is "nothing but" changing brain states. With this concept Berkeley and Hume built tables out of squareness and brownness, Pavlov built intellect out of reflexes, and the physiologist Jacques Loeb built life out of tropisms.[1] As long as the universe is seen as continuous, this "error of origins" can be perceived as a reasonable deduction. But as we have also seen, the idea of a consistently continuous universe with no leaps was destroyed by Max Planck in 1900, and it is no longer a basis of modern science. Looking at it from today's viewpoint, we can see that it never was a "rule of nature" (whatever that phrase means), but at best a useful convention of thought that is now, for our purposes, a nuisance giving rise to unnecessary problems. Once we insist that the data in each domain must be taken on their own terms and not squeezed into the Procrustean bed of the sensory realm, it becomes clear that it is both valid and scientifically necessary to view honor as qualitatively different from cultural conditioning, love as more than an itching in the groin, and consciousness more than brain structure and physiology.

We must now abandon the chicken-and-egg conundrum and give up worrying about what is basic to, and causes, what. After all, as Joseph Wood Krutch has pointed out, it is as reasonable to say that genital urges are the simplest and crudest manifestations of a fundamental reality called "love," as the other way around.[2] Neither proposition is more philosophically secure than the other, and each has advantages and disadvantages. One is reminded here of the old Spanish proverb quoted by the philosopher Antony Flew: "Take what you want," said God, "take it and pay for it!" However, from the viewpoint of modern science, we should take neither of the above propositions nor the viewpoints they embody. They belong to different domains, and each of them must be dealt with on its own terms. Neither proposition is an epiphenomenon of the domain of the other. The brain no more secretes consciousness than consciousness secretes the brain. Different domains have different observables.

The greatest insight of the present stage of the evolution of knowledge (roughly from the 1880s to the present) is that if data do not fit (behave lawfully in) the accepted, everyday organization of reality ("metaphysical system," "definition of what is and what is not," etc.), then it is necessary to reorder and reorganize the concept of reality from which the data came, so that they do fit it, do behave lawfully within it. That is, if the data are not lawful in the ordinary way of organizing reality, then they must be taken as facts, and the organization of reality as theory and the theory must bow to the facts.[3] This, however, is not an all-or-none phenomenon: The concept of reality must only be changed in the realm in which the data are not lawful, and kept in the realms in which they are lawful. It is this insight that the social sciences have used only to a very limited degree. Most commonly, the only data that have been accepted as useful and valid, worked with and not ignored, were those that fit the ordinary Western "commonsense" definition of reality.

As we have tried to indicate, matters that seem "obvious" to us, matters that our intuition tells us are clearly true, true without need of proof, are generally true only in the see-touch realm. However, when we go beyond this realm, our intuition about how things are and work is frequently false; the "obvious" is no longer necessarily valid. To take a few diverse examples: Our intuition would never tell us (and, indeed, contradicts the fact) that there are as many even numbers as there are even and odd numbers combined; that things increase in mass as they move faster; or that the more we know about the position of an *on*, the less we can know about its velocity, and vice versa. In the realm to which these data belong our intuition does not work.

In the words of the mathematician and philosopher L. V. Bertalanffy

... in recent times ... the Kantian categories—supposedly eternal *a prioris* for any thinking being—had to go because all too human and fitting only the familiar world of medium dimensions, but not the worlds of the immensely small and immensely big which came into the field of scientific research.[4]

However, we must be careful not to throw out the baby with the bathwater. Our intuition *does* work in the sensory realm. When V.V.O. Quine says, echoing other modern philosophers, "Intuition is bankrupt," he is overstating the case. It is only bankrupt in domains where a mechanical model does not apply. One of these domains is the domain of consciousness.

In an elaborate analogy in which he compared the present existence and the history of the city of Rome to the mind, Freud demonstrated that it "leads to the inconceivable, or even to absurdities, to try to master the idiosyncrasies of mental life ... by treating them in terms of visual representation."[5] It is, he wrote "... impossible to represent phenomena of this kind in visual terms."[6]

We cannot, as we have indicated before, *quantify* the observables in the domain of consciousness. There are no rules of correspondence possible that would enable us to quantify our feelings. We can make statements of the *relative* intensity of feelings, but we cannot go beyond this. I can say, "I feel angrier at him today than I did yesterday." We cannot, however, make meaningful statements such as, "I feel three and one half times angrier than I did yesterday."

Because of this lack of rules of correspondence, we cannot really compare the intensity of feelings of one person with the intensity of feelings of another. To say "My joy is greater than yours," is as meaningless as to say (in Wittgenstein's phrase), "I have a pain in your leg." A story told by the psychiatrist Victor Frankl illustrates this idea. A Prussian officer and an Austrian private by chance found themselves in the same shellhole in a bombardment in World War I. The Prussian officer asked, "Are you afraid?" and the private replied, "I am very, very frightened." The Prussian said, "This illustrates the superiority of race and training. *I* am not afraid." The Austrian private answered, "It illustrates the difference, but not the superiority.

If you were one half as frightened as I am, you would have run away long ago!"

As far back as 1926 the physicist, A. S. Eddington pointed out that consciousness has two aspects. The first is as a component in the organization of reality as it is perceived. ("Man is the being as a result of whose appearance a world exists," wrote Sartre.)[7] The second aspect, said Eddington, is self-awareness. He believed that the first function is quantitative (at least theoretically) and the second function is nonquantitative (even theoretically). It is this Eddington believed, that is the important difference between the two.

... the division of the external world into a material world and a spiritual world is superficial and the deep line of cleavage is between the metrical and the non-metrical aspects of the world. . . . The "likeableness" of a particular elephant is non-metrical, its "ponderosity" is metrical.[8]

We believe this to be an important insight although the matter now appears to be more complex than Eddington's statements would indicate. For example, we might point out that as human beings become aware of the second aspect, self-awareness, they can devise methods to change the self. These would include education, meditation, psychotherapy, and so forth. As we become aware of the first function, the construction and organization of reality, we can examine the methods by which we attempt this organization, find the flaws in our methodologies, and devise new ones. We believe that up to the present period, this option for change has not been possible because we were not aware of the very great degree to which we constructed reality. The heart of the very rapid evolution of science and of the "age of transition" in which we live is the growing awareness of how we invent as much as discover reality.

As we have indicated in chapter 12 social scientists have to a very large degree modeled their work on that of the successful science of the 18th and 19th centuries. Illustrating the physicist Max Born's statement, "The physics of one period is the metaphysics of the next," they have projected into all realms of experience the pattern of being found successful in the sensory realm, deciding in advance that the data they would look for and accept as valid were the data that could be assimilated into the machine model of the sensory realm. And as we have also noted earlier, Cartesian Dualism, the formulation of reality used in 18th- and 19th-century physics, is designed for the study of only one-half of the dualism, the "outer world," the *res extensa,* and is not designed for the study of consciousness, the other half, the *res cogitans.* For extending our comprehension of this realm, it proved largely useless.

The physicists' schema, so faithfully emulated by generations of psychologists, epistemologists and aestheticians, is probably blocking their progress, defeating possible insights by its prejudicial force. The schema is not false—it is perfectly reasonable—but it is bootless for the study of mental phenomena.[9]

With no rules of correspondence possible and no ability to quantify the data, what *can* we say about the observables of the domain of consciousness? We can formulate some general statements:

1. The observables will be different in sharpness and distinguishability from those in the sensory realm. There are no "things" in this realm, only "processes." (This well-known fact derives both from experience and from attempts to analyze these observables. It is not, however universally accepted, because of the prevalent belief that our inability to discover 'things' in this realm is due to our ignorance, and that if we could analyze the data correctly, the observables would turn out to be similar to those found in the see-touch realm. Again, the data simply have not been taken in their own terms.)[10]

2. Observables have "limited access"—that is, they can only be observed by one person—in contrast to the "public access" of observables in many other realms.

3. The observables cannot be quantified in an absolute manner. There are no rules of correspondence: We cannot—in principle— say, "He will feel three exuberants of joy when he hears the news."

4. Since the observables cannot be seen or touched, they do not have many of the characteristics of observables of the sensory realm. They have no size ("how large is a fear?"). They cannot be localized in space ("where do I feel love?"). They have no shape, surface, or color.

5. Since the observables cannot be quantified and cannot be located in space, their interaction will be different from the interaction of observables in the sensory realm. The basically simple algebraic principles governing interaction of sensory realm observables will simply not be applicable in the domain of consciousness.

6. The guiding principles of space and time in this realm are different from those of the sensory realm. Space here is personal space, time is personal time. These definitions, needed to make the data lawful, are not the same as Euclidean (yardstick) space or Newtonian

(clock) time. There is no reason to suppose that they are the same as the space-time of the macrocosm. So far no mathematical systems have been shown to be applicable to personal space and personal time.[11]

7. There are no conservation principles. (As, "matter and energy are neither created nor destroyed.") We expect miracles of creation when a genius produces new ideas. These ideas neither violate nor obey the conservation laws found in many other domains; they simply "happen."[12]

8. The Second Law of Thermodynamics is not applicable. Mental activity can increase order in its own domain (as by drawing new conclusions from previously known facts); this is not done at the expense of a larger environment.

9. Mechanical models are not valid. The existence and interaction of observables cannot be visualized according to the push-pull principles of a machine.

10. Both purposes and causes exist as observables. Free will is an observable with no counterpart among the observables of the sensory realm. Will is oriented to its future. It may well be that future work will determine that free will—an observable in the domain of consciousness and of the molar behavior of the individual human being—is not an observable in the domain of the molar behavior of large numbers of human beings. We can confidently predict that, barring the occurrence of an earthquake or hurricane, a very large number of people will be on the George Washington Bridge leaving New York City next Friday at 5:30 P.M. But you and I may decide to stay in town that night and go to the theatre. Mass behavior may be predictable, although this is far from proven. Individual behavior is not. This, of course, is not a new idea. We see, for example, the same kind of differentiation between the microcosm and the sensory realm. The unpredictability and "statistical causation" of the realm of onta becomes "specific event predictability" when such large numbers of onta are involved that we are in the sensory realm. No paradox is involved. Similarly "free will" in the realm of one human being may well become "statistical predictability" or even "specific event predictability" in the realm of large numbers of human beings. Although this is speculation at the present stage of our knowledge, it seems likely that the domains of the consciousness and behavior of the individual human being are in different *realms* than the domain of the behavior of large numbers of human beings.

Absolute prediction of future specific events is not possible, but statistical prediction—in a relativistic sense—is possible. We can say, "She is more likely to feel joy than feel sorrow when she hears this news." We cannot say, "She *will* feel joy," or "There is a thirty-three percent chance of her feeling joy."

Causation (in the sense of predictability of specific events) works only for isolated systems. In the physical world this isolation can often be brought about. In the domain of consciousness it cannot. If you institute the rigorous procedures needed to begin to isolate a mind (such as sensory-deprivation procedures), you first damage the mind, then break it down. The mind *resists* isolation attempts. Consciousness constantly reaches to the past and the future, to other places and possibilities of places.

11. Future conscious events are unpredictable in principle. Past events can be shown to have been determined. From a scientific viewpoint, then, the past of an individual consciousness is determined, and the future is free.

The acceptance of the mechanical view of man has gradually increased over the past two centuries. Neither the social sciences nor academic philosophy have made headway against it. As we have noted, only from modern physics did there appear the possibility of a way out. First, physics showed the necessity of the concept of different realms of experience having different (but compatible) metaphysical systems necessary to make their data lawful; second, modern physics provided a scientific basis for free will. It is the latter that pertains now to our discussion of the domain of consciousness.

In chapters 7, 8, and 10 we explored the fallacy of any reductionism that attempts to explain every aspect of physics and other sciences in terms of matter using the pre-quantum view.[13] In a wider sense we introduced the need for the principle of transcendence with compatibility, the need to invoke agencies whose essence cannot be foreseen and is often meaningless in other scientific domains. We demonstrated that the mind-body problem cannot be resolved by any reductive act, thus disposing of identity theories and of epiphenomenalism, monism, dualism, biperspectivism, and similar doctrines. The mind, we concluded, is *sui generis;* it transcends the body but can nevertheless interact with it.

Our conception of the mind differs in one important respect from others proposed chiefly by parapsychologists and vitalists.[14] We do not assign to the mind such physical attributes as force—thus denying vitalism—and energy (the mind is not a physical agency, so attributes such as energy are not likely to be among its observables. We do assume interactions between mind and brain but none involving a transfer of energy.

No one will deny that an interaction between mind and body takes place whenever we consciously perform a movement. We now make the additional affirmation that our will—the core of consciousness, wherein the self proclaims its being most emphatically—interacts with the body in a special way when it makes a decision and deliberately activates the body. In pre-quantum days, when philosophy was dominated by Laplacian determinism, in which a state classically defined without recourse to probabilities rigorously entailed all future states (of an isolated system), free will was a paradox and an illusion. That is to say, either it could not be explained, despite the immediate, empirically accurate evidence that affirmed it, or its affirmation was false. This situation has changed by virtue of the discovery of quantum mechanics. The new discipline provides the possibility of a solution by removing the impediment of old-style determinism.

Our thesis is that quantum mechanics leaves our body, our brain, at any moment in a state with numerous (because of its complexity we might say *innumerable*) possible futures, each with a predetermined probability. Freedom involves two components: *chance* (existence of a genuine set of alternatives) and *choice*. Quantum mechanics provides the chance, and we shall argue that only the mind can make the choice by selecting (not energetically enforcing) among the possible future courses.

But first we must remove an obstacle. It will be recalled that probabilities of alternate courses of events rule the microcosm, and that the probabilities congeal to certainties in the molar world to which the human body belongs. It might seem, therefore, that neural processes are subject to strict, or nearly strict, determinism. The criterion for determining whether chance prevails is Heisenberg's *Uncertainty Relation* which we consider again briefly in the form

$$\Delta x \cdot \Delta v \geqslant \frac{\hbar}{2m}$$

Here Δx and Δv are ranges of position and velocity of an *on* having mass m, ranges over which probabilities have control. They are usually called, somewhat loosely, errors of observation. Clearly the value of m is critical. The symbol $\hbar$ is Planck's constant divided by 2π.[15]

Let us therefore consider what masses are involved in brain processes and in the excitation of our sensory organs. To get our initial bearing, we first apply Heisenberg's *Uncertainty Relation* to the electron of the hydrogen atom, which we have already discussed. Its mass is such that $\frac{\hbar}{2m}$ is about one (in centimeter, gram, second units). We recorded previously that the electron's speed around its smallest orbit is about $\pm\ 10^8$ centimeters per second (the plus/minus sign is to indicate that it is sometimes to the right, sometimes to the left).[16] This results in Δx being approximately 10^{-8} centimeters, which is about the atom's size, i.e., the distance within which its position is uncertain. For an entire brain cell the mass is at least one trillion times as great as it is for an electron; if we allow for Δx an uncertainty of measurable size—say, one millimeter—Δv becomes 10^{-11} centimeters per second, a speed range so small that it could hardly be responsible for any physiological action. The mass of a synaptic knob within the cortex is even greater and therefore precludes any uncertainties, any probability concerns, still more effectively. For them determinism holds.[17]

But this is hardly relevant in the present context, for who is interested in the motion of the brain cell or a neuron as a whole? What triggers neural impulses is more likely to be a single atom or molecule, or perhaps several of them. Their masses are about 100,000 times that of an electron, and if the intrinsic uncertainty of their position were of the dimension of a brain cell, the uncertainty in their velocities could be centimeters per second.

The distinguished neurophysiologist and philosopher J. C. Eccles[18] considered the possibility that onta within cells and synaptic knobs, perhaps even electrons, which are free under physical indeterminacy, may have their behavior influenced by the mind. And if an electron is confined in a brain cell, its velocity would range from zero to 10^6 centimeters per second, which is about the speed of an intercontinental ballistic missile. If the mind can choose from this enormous range of velocities to trigger some physiological process leading to bodily action, freedom of the will is no longer a paradox.

A relation similar to (1) also holds for energy and time; it reads $\Delta E \cdot \Delta t \geqslant \frac{\hbar}{2m}$, where E means energy, t time, and Δ, as before, "uncertainty in." Precisely what is meant here by delta t (Δt), the "uncertainty" in the time that a physical process involving a mass m requires, is somewhat obscure in the present context (and in many physical problems to which this inequality is applied), but whatever reasonable value we choose for Δt in connection with electron or ion motion within a cell, we obtain a value for E—the energy uncertainty from which the mind can choose—that is much larger than the energy that can stimulate the mind to its awareness. The latter is only about 10^{-11} ergs for the sensation of hearing! For vision the energy of a single photon (10^{-12} ergs—one one hundred billionth of the customary small unit of energy) can trigger a response.

Seeing provides another convincing example. Our retina is sensitive to single photons. Here we need hardly appeal to the uncertainty principle for numerical evidence, for we know that a photon is subject to the laws, and displays the probabilistic behavior, of quantum onta.

The preceding considerations offer a solution to the free will problem: physics provides genuine alternatives for choice, and the mind chooses. Along the way we have been forced to assume that the mind, as an independent entity, can perform that selection—in other words, can influence the body. We shall henceforth accept this as a postulate. To summarize our findings with respect to freedom of the will, we repeat here in part the conclusion of an earlier paper in which this solution of the free will problem was proposed.[19]

Classical determinism made freedom intrinsically impossible unless its application to psychophysical phenomena was arbitrarily excluded.

Historical arguments designed to reconcile freedom with classical causality were able merely to establish a subjective illusion, a personal feeling of freedom.

Modern physics, through Heisenberg's principle of indeterminacy, has loosened Laplacian determinism sufficiently to allow for *uncaused* atomic events, permitting in certain specifiable situations the incidence of genuine chance.

The consequences of such microcosmic indeterminacies, while usually insignificant in the molar world, do ingress into the macrocosm (at least in several known instances), and even play a role in delicate neurophysical and chemical processes, and these have major effects on behavior.

Physics thus makes understandable the occurrence of *chance*, of true alternatives upon which the course of events must seize. Physics alone, in its present state, *can* account for unpredictable, erratic human behavior.

Human freedom involves more than chance: It joins chance with deliberate choice. But it needs the chance. So long as science can say nothing about this active, decisive, creative element called choice, it has not fully solved the problem of freedom. Our proposed solution is simple. We invoke mind, or consciousness, as the independent agency effecting the choice. In historical retrospect we thus fulfill Augustine's prescription in *De Libero Arbitrio*. He saw that human freedom involves two ingredients: chance and choice. He was able to identify the choice but not the chance. Now quantum theory has come to his rescue.

And now an afterthought. We have suggested that a nonphysical agency chooses among physical possibilities. Would this fuller understanding not restore determinism? If we can explain how the agency effecting choice selects from the alternatives presented by physics, will the inclusion of that agency in the scheme of things not leave us where we started—i.e., with an amplified Laplacian formula?

The answer is negative, for that agency is the mind, and we know it to be teleological in its function. It looks into the future rather than into the past, is drawn by purposes as well as impelled by drives, partakes of the liveliness of the incalculable human spirit—freedom in a unique sense survives.

Notes

CHAPTER 1

1. A. Koestler, *The Act of Creation.* London: PAN, 1970, p. 253.
2. Giordano Bruno wrote: "God is not an external intelligence, rolling around and leading around; it is more worthy of him to be the internal principle of motion, which is his own
nature, his own appearance, his own soul . . ." Quoted in E. Cassirer, *The Philosophy of the
Enlightenment.* Princeton, N.J.: Princeton University Press, 1951, p. 41.
"The general laws of nature which govern and determine all phenomena are nothing but
the eternal decrees of God which always entail eternal truth and necessity." Spinoza, *Trac-
tatus Theologico.* Politicus, III, 7.
3. "The aspiration to demonstrate that the universe ran like a piece of clockwork . . . was
itself initially a religious aspiration. It was felt that there would be something defective in
creation itself—something not quite worthy of God—unless the whole pattern of the universe could be shown to be interlocking, so that it carried the pattern of reasonableness,
and orderliness." H. Butterfield, *The Origins of Modern Science.* London: G. Bell, 1949, p.
105.
Because of this concept of one rationality governing the entire universe, the Western
world rejected the irrational as a way of gaining truth. The intellectual thought of the West
rejected the strong and clear stream of Western mysticism and ultimately came to the conclusion that this approach led away from, rather than toward, truth.
"The highest energy and deepest truth of the mind [in this view] do not consist in going
out into the infinite, but in the mind maintaining itself against the infinite, and providing in
its pure unity equal to the infinity of being. Giordano Bruno, in whom this new climate of
opinion first appears, defines the relation between the ego and the world, between subject
and object in this sense . . . . The power of reason is our only access to the infinite, it assures
us of the presence of the infinite and teaches us to place it within measure and bound, not
in order to limit its realm, but in order to know it in its all-comprehensive and all-pervasive
law." Cassirer, op. cit., p. 38.
Even before Bruno, there was never any conflict perceived in the West concerning the
results of different approaches to truth. Aquinas believed that if the twin approaches of revelation and faith on the one hand, and observation and reason on the other, were both applied to the same problem, and were each used correctly, they would come to the same
conclusion. The great scholastic schools saw as their main task to harmonize and reconcile

faith and reason, belief and knowledge. From their viewpoint, there could never be a real contradiction between these two.

In the East, it was quite otherwise than in the West after Bruno. Eastern intellectual and mystical thought were integrated and seen as a valid synthesis. This, of course, had disadvantages as well as advantages. Jung put it: "The Hindus are notoriously weak in rational exposition. They think for the most part in parables or images. They are not interested in reason. That of course is a basic condition of the Orient as a whole." C. Serrano, *C.G. Jung and Herman Hesse: A Record of Two Friendships*. New York: Schocken, 1916, p. 50.

4. In a fascinating poem, *The Masque of Reason*, the poet Robert Frost explains the treatment of Job as God's attempt ". . . to stultify the Deuteronomist": to attain freedom from the inexorable laws of action and response that He had created and was now trapped by."

5. "Galileo, Descartes, and Newton all regarded God as a kind of 'chief mathematician' of the Universe. 'Geometry existed before the creation, is coeternal with the mind of God, is God himself,' wrote Kepler and the other giants echoed his conviction." Koestler, op. cit., p. 262.

"The scientist knows that there are still very large fields of phenomena which it has not yet found possible to reduce to strict laws and to exact numerical rules. Nevertheless he remains faithful to the general Pythagorean creed: he thinks that nature taken as a whole and in all its specific fields is "a number and a harmony." E. Cassirer, *An Essay on Man*. New York: Anchor Books, 1954, p. 277.

6. Leaving aside the occasional oddities like two raindrops merging and making one raindrop. Even in these, however, the quantity of liquid doubled when the number of raindrops halved so the quantitative view was generally salvageable and viable. If, for example, you could not see at first where the energy of a coiled spring went when it was dissolved in acid, you presently found out that it went into heating the acid more than if the spring were uncoiled. Matter and energy could be counted upon to remain constant if you looked carefully enough; they did not appear and disappear.

7. To an attack on this concept made by quantum mechanics, Einstein replied, "I cannot believe that God plays dice with the Universe."

8. "When an observation is made on any atomic system that has been prepared in a given way and is thus in a given state, the result will not in general be determinate—i.e., if the experiment is repeated several times under identical conditions, several different results may be obtained." P.A.M. Dirac. Quoted in J. Jeans, *The New Background of Science*, 1934, p. 47. For further details see chapter 11 of this book.

9. In William James' terms the statement "All crows are black" becomes false as soon as you see one white crow.

10. "If one wants to give an accurate description of the elementary particle—and here the emphasis is on the word 'accurate'—the only thing that can be written down as description is a probability function. But then one sees that not even the quality of being (if that may be called a 'quality') belongs to what is described. It is a possibility for being or a tendency for being." W. Heisenberg, *Physics and Philosophy*. New York: Harper, 1958, p. 70.

11. H. Margenau, *Open Vistas: Philosophical Perspectives of Modern Science*. New Haven: Yale University Press, 1961.

Our use of the word "on" requires an explanation—perhaps an apology. It was introduced by one of the authors on an earlier occasion but produced little in the way of echoes. It is the present participle of the Greek verb εἰναι, "to be." Hence it means *being* in its largest sense. The reason for the choice is this: In the quantum theory we encounter numerous "elementary particles." But it is clear today that they are not particles at all, nor are they waves or vortices or electrical point charges or anything that can be defined in terms of visual characteristics. They simply transcend the molar, the see-touch realm. Most of their names end in "on," the suffix of many Greek nouns. For this reason we desire to use the most abstract, brief expression available, and so we selected "on." Its literal meaning is the same as the Latin ens, from which the term "entity" is derived. Hence "on" and "entity" are synonyms, and we selected the former because its novelty matches the novelty that must be recognized in the elementary "particles." The plural of "on" is onta.

Physicists are known to be fussy, and perhaps fuzzy, in their choice of terms. We recall an episode that occurred at the time at which the positron, the positive electron, was discovered. The secretary of the American Physical Society, Karl Darrow, a student of Greek literature, suggested the term "Oreston" after Orestes, the half-brother of Electra. This, according to a personal communication from Darrow, was rejected by prurient members of the Society who knew of the classical rumor of incest between Electra and Orestes. In afterthought the interaction between electrons and positrons might seem to justify that ancient rumor in terms of what happens when they meet. They destroy each other and give birth to photons.

12. In Einstein's words: "Before Clerk Maxwell, people conceived of physical reality—insofar as it is supposed to represent events in nature—as material points, whose changes consist exclusively of motions . . . after Maxwell, they conceived of physical reality as represented by continuous fields, not mechanically explicable. . . . This change in the concept of reality is the most profound and fruitful one that has come to physics since Newton." Quoted in H. Margenau, "Einstein's Conception of Reality." In P. A. Schilpp, ed., *Albert Einstein: Philosopher Scientist.* New York: Harper, 1959, p. 253.

13. These were attempts to make visual or mechanical models of the parameters of our experience of taste and smell. None proved particularly useful, and none gave a sense of correctness or recognition.

14. The last two paragraphs are a paraphrase of some remarks of Peter Berger in his *Rumor of Angels.* New York: Doubleday, 1961, p. 71 ff.

15. There are certainly major exceptions in all these fields—such as Peter Berger in sociology, Jacob Needleman in philosophy, and Carl Rogers in psychology—but they are far from the rule. Another major exception is Ernst Cassirer. In his *Language and Myth* he deals with the two kinds of thinking he saw in the structure of language and described each as ideally suited for its purpose. There is certainly no sense in which either is superior to the other.

16. "Where id was, there shall ego be. It is reclamation work, like the draining of the Zuyder Zee." S. Freud, *New Introductory Lectures.* New York: Norton, 1933, p. 112.

17. Piaget has developed these methods to a very high degree in his work with young children.

18. What, for example, of Carlos Casteneda's description of one construction of Reality? (*A Separate Reality.* New York: Pocket Books, 1972, and his other books on this subject.) Is it a valid or a schizophrenic one? We personally believe the latter, although we would be hard put to prove it.

19. A. Eddington, *The Philosophy of Physical Science.* Ann Arbor: Michigan University Press, 1959, p. 31.

20. In chess, the move "King's Knight to Queen 3" tells us only of two locations of the knight. It tells us nothing of what went on between its being in one and then the other of these two locations. We can have no idea of the "route" it followed from one to the other. The rules of the game make no reference to a knight being between locations either spatially or temporally. It was there, is here, is affected by nothing, and affects nothing in between. Further, there is no time lapse—its "movement" simply does not take any time at all to accomplish. It literally is never "in-between."

The universe of chess is a limited, but perfectly coherent world. It has a well-organized metaphysical system that interacts with its "reality," which includes the two-dimensional board, the limited number of pieces, etc. This includes such Basic Limiting Principles as the impossibility of two pieces occupying the same space at the same time, the piece in motion always being superior to the piece at rest, and so forth.

21. In the theory of the Compton effect (collision between an electron and a photon) the elementary treatment does use the laws of conservation of energy and momentum as though these onta were billiard balls. They are indeed satisfied, but the directions after impact could not be predicted.

22. A. Einstein and L. Infeld, *The Evolution of Physics*. New York: Simon and Schuster, 1961.

". . . a really relativistic physics cannot be based on Euclidean geometry." (p. 266)

". . . if we wish to reject absolute motion and keep up the idea of the general theory of relativity, then physics must be built on the basis of geometry more general than the Euclidean." (p. 228)

23. In the see-touch realm, for example, mass, velocity, and time are considered separately. Thus we say that an arrow with a certain velocity and a certain mass will travel a certain distance in a certain amount of time. Velocity, mass, and time are related but independent. In the realm of the too large or too fast, however, quite a different situation exists. The greater the velocity of our arrow, the greater its mass becomes and the slower runs the clocks of any bacteria riding along on it. (Also, it becomes shorter.)

The two ways of organizing reality in these two realms are completely compatible. Although they are very different, they are not contradictory. It is legitimate to say that the construction used in the see-touch world is a "special case" of the construction of reality used in the too large or too fast world. Nevertheless, they are very different constructions of reality.

24. R. Abel, *Man is the Measure*. New York: Free Press, 1976, p. 270.

25. W. Heisenberg, *Philosophical Problems of Nuclear Science*. New York: Fawcett, 1966, p. 13.

CHAPTER 2

1. Abel, *Man Is the Measure*.

2. N. Rashevsky, "Is the concept of the organism as a machine a useful one?" In P. G. Frank, ed., *The Validation of Scientific Theories*, Boston: Beacon, 1954.

3. E. Underhill, *Practical Mysticism*. London: J. M. Dent, 1964, p. 27.

4. "The fact that we can never *know* what reality is, is clarified by the following. A guess is made as to how things are or work. We wish to compare it to 'the truth.' What does this mean? We cannot even guess. What a comparison would mean is completely unclear." A. Einstein and L. Infeld, *The Evolution of Physics*. New York: Simon and Schuster, 1938, p. 35.

5. E. Schrödinger, *Mind and Matter*. London: Cambridge University Press, 1959, p. 52.

6. H. Smith, *Condemned to Meaning*. New York: Harper, 1956, p. 43.

7. "If it is true that every theory must be based upon observed facts, it is equally true that facts cannot be observed without the guidance of some theory. Without some guidance, our facts would be desultory and fruitless; we could not retain them: *for the most part we could not even perceive them.*" (Italics ours) A. Comte, "The Positive Philosophy." In S. Commins and R. Linscott, eds., *Man and the Universe: The Philosophers of Science*. New York: Washington Square Press, 1954, p. 225.

8. B. Russell. Quoted in A. Korzybski, *Science and Sanity*, 3rd ed. Lancaster, Penn.: Non-Aristotelian Publishing Co., 1933, p. 263.

9. S. Langer, "On Cassirer's theory of language and myth." In P. A. Schilpp. ed., *The Philosophy of Ernst Cassirer*. Evanston, Ill.: Library of Living Philosophers, 1949, p. 381.

10. K. R. Popper and J. C. Eccles, *The Self and Its Brain*. Berlin: Springer International, 1977, p. 61.

11. "By convention colored, by convention sweet, by convention bitter. In reality only atoms and the void." Democritus. These examples have been largely drawn from Abel, op. cit.

12. Abel. op. cit., p. 215.

13. Several of these examples are from Abel op. cit., p. 106.

14. In Jung's words, during the Medieval period "Men were all children of God under the loving care of the Most High who prepared them for eternal blessedness." There is a very great difference between being raised and living in a culture where this is seen as common sense, and being raised and living in a culture where common sense indicated rather that each is a completely replaceable cog in a great machine that simply mindlessly goes on. C.G. Jung, *Modern Man in Search of a Soul.* London: Methuen, 1953, p. 233.

15. "When Descartes drew a sharp line between matter and spirit, he at the same time drew a line between scientists and philosophers. Henceforward scientists took charge of matter and philosophers dealt as best they could with spirit." T. Walker, *The Diagnosis of Man,* Baltimore: Penguin, 1962, p. 152.

This was the first time that this had been done. In the past there was no separation of scientists and philosophers. The split, the last evidence of which was the separation of academic psychology and philosophy departments, must be healed before real progress can be made. The physicists who made the great advances past the see-touch realm were as much philosophers as scientists and one cannot separate their physics from their metaphysics.

16. L. LeShan, *Alternate Realities.* New York: Evans, 1975.

17. We shall discuss this in more detail in chapters 7, 8, and 10.

18. We feel that the hope of an establishment of a unique connection between all states of depression and brain states is illusory, reminding the physicist of the frustrated search for hidden variables in quantum mechanics.

CHAPTER 4

1. An appeal to randomness, so often made by scientists steeped in the epistemology of physical reality, always suffers from the circumstance that we possess no clear and generally accepted definition of the term, nor an absolute criterion for a random event.

CHAPTER 6

1. The caloric theory took heat to be an imponderable fluid. The kinetic theory took it to be the kinetic energy of the molecules encompassing a body.

2. In 1798 Count Rumford (An American Tory who fled to Europe during the revolution) was boring cannon for the elector of Bohemia. On observing the large amounts of heat generated during the boring operation, he decided that heat must be a form of motion.

3. That is, if you slowly compress a gas in a container equipped with a pressure gauge and a device measuring the volume.

4. For the definition of "probable error" see R. B. Lindsay and H. Margenau, *Foundations of Physics.* New York: Dover, 1939, or any textbook on the theory of measurement.

5. This is sometimes risky because it involves a number called the a priori probability of success. This is difficult to ascertain and has a crucial effect on the results. The formula itself is given, for example, in H. Margenau and G. M. Murphy, *The Mathematics of Physics and Chemistry.* New York: D. Van Nostrand, 1943.

6. Mrs. Garratt occasionally invited one of the authors to meetings by parapsychologists in her beautiful home in Le Piol, hoping that he would achieve such coordination. Unfortunately, he did not succeed, but he blames his failure on his preoccupation with other tasks. In chapter 15 we give some suggestions that might lead to the desired coordination.

CHAPTER 7

1. A. Coleman, *Trans. Royal Astr. Soc.,* Vol. 1, 1930.

2. This might loosely be called the first attempt at a unified field theory.

3. For a more detailed analysis, see also H. Margenau, *The Nature of Physical Reality*. New York: McGraw-Hill, 1950.

4. Chapter 4.

5. Cf. E. Laszlo, *Introduction to Systems Philosophy*. New York: Gordon and Breach, 1972.

6. Similar arguments will be found in M. Bunge, "The Metaphysics, Epistemology and Methodology of Levels." In White, Wilson, and Wilson, eds., *Hierarchical Structures*. New York: Elsevier, 1969. Approaches to the view we present appear in H. Margenau, ed., *Main Principles of Modern Thought*, chap. 2. New York: Gordon and Breach, 1974, and in Laszlo, op. cit.

7. More fully in Margenau, op. cit.

8. The preceding account represents the standard view. An interesting objection to it, which has apparently not found its way into the literature, has been voiced by Max Born, the famous quantum theorist, in a private conversation. He points out that perfectly exact knowledge of the position and velocity of each gas molecule, or even a single particle, is in principle impossible to obtain. (We shall return to this point later when discussing epistemic feedback.) The human mind must always reckon with errors of observation. Now a computation shows the following. If a very small error in our present knowledge of the position of the molecules of air in a normal room were admitted and taken into account in the calculation, an application of Newton's laws would not preserve the accuracy of the initial knowledge. It happens that every collision between two molecules increases the error. For instance, if the initial error amounted to only about .005% it would grow to 100% in less than a microsecond; the original knowledge would have become total ignorance.

This consideration, while highly interesting, does of course not alter what was said about the failure of reciprocal reducibility between Newtonian and statistical mechanics.

9. For further details see R.B. Lindsay, and H. Margenau, *Foundations of Physics*. New Haven, Conn.: Oxbow Press, 1980.

10. To be specific: there is, besides four distinct fundamental fields, a confusing multitude of so-called elementary particles, bearing observables like mass, charge, spin, isospin, strangeness—even charm, color and flavor (in a highly metaphoric sense).

CHAPTER 8

1. To be sure these have meaning only if certain restrictions are placed upon the A observables (Maxwellian distribution), but in a short time these restrictions realize themselves automatically.

2. This argument has also been given in H. Margenau, *Bunge Festschrift*.

3. The term emerging, frequently used by the advocates of reductionism, does not quite fit our examples and our view. For what "emerges" was previously present but invisible, while we wish to emphasize the novelty of new constructs and observables, their being created in the passage to another mode of explanation.

4. We use the term "molar" in its customary sense, as distinct from atomic and molecular, but also distinct from the astronomically large. "Ordinary" or "accessible to the senses" might be used to replace it.

5. It can be characterized most simply by saying: It is the space in which the Pythagorean theorem, i.e., the sum of the squares of two sides of a right triangle equals the square of the hypotenuse, is true.

6. See, for instance, C. W. Misner, K. S. Thorne and J. A. Wheeler, *Gravitation*. San Francisco: Freeman Press, 1973.

7. P. Jordan, *Die Herkunft der Sterne, Wiss*. Verl. Gesellschaft, 1947.

8. Margenau, *Open Vistas*.

For the reader who is interested in a little more detail we add the following. Draw a Cartesian (rectangular) set of coordinates in a plane and label them x and y. Any point, P, in the plane has a distance s from the origin, which, according to Pythagoras, satisfies the relation $S^2 = x^2 + y^2$. In three dimensions, when we add a z-axis, this would read $s^2 = x^2 + y^2 + z^2$. Now consider two different points in this three-space, but let them be very close together. When these two points are projected upon the x-axis, the distance between these projections is dx, and in a similar way we construct and define dy and dz. Then, again according to Pythagoras, the distance between the two neighboring points, designated as ds, satisfies $ds^2 = dx^2 + dy^2 + dz^2$.

This equation, called the metric equation, is characteristic of Euclidean geometry. In non-Euclidean geometry it is replaced by $ds^2 = adx^2 + bdy^2 + cdz^2$, where a, b, c are numbers different from 1, but in all present physical applications they are very close to 1. These geometries describe spaces that are curved, which is another way of saying that the shortest distance between two points may be a curve and not a straight line.

Finally, we encounter the following remarkable result. If, instead of dealing with two points in three-dimensional space, we consider two events that are close together in both space and time, their space coordinates being dx, dy, and dz as before, but the small interval of time that separates them is dt, then according to Einstein one introduces a four-dimensional interval ds, which obeys the equation: $ds^2 = dx^2 + dy^2 - c^2 dt^2$.

This equation defines the sense in which time is said to be a fourth coordinate, a fourth dimension of space. It is fundamental in the theory of special relativity. The constant c is the velocity of light.

9. They are, however, related to experiment (P-plane) by indirect and complex rules of correspondence.

10. Margenau, *Philosophy of Science*, 1949. pp. 16, 287.

11. Schrödinger, in a private communication, suggested that the notion of latency should be applied even to the identity of elementary particles—indeed, even to their existence. His conjecture was that they could be called into being the act of measurement. This has been found to be true.

12. The various E_i may form a partly discrete and partly continuous sequence of values.

13. There is already a tendency on the part of some nuclear physicists to interpret recently found elementary particles of unusually great mass as higher quantum states of lighter particles.

14. This term is used by E. Laszlo to characterize the mind-brain relation. See his introduction to *Systems Philosophy*. New York: Gordon and Breach, 1972.

15. The physicists' use of the word "classical" in this context is peculiar; while to the humanist it means excellent, time-honored, worthy of admiration, the physicist applies it to a theory that has lost its ultimate validity.

CHAPTER 9

1. The biologist Theodore von Uëxkull has explored this matter and—from his knowledge of comparative anatomy—has drawn pictures of how the same scene would appear to a variety of different species. See, for example his "A Walk in the Worlds of Animals and Men." In C. H. Schiller, tr. and ed., *Instinctive Behavior*, New York: International Universities Press, 1957.

CHAPTER 10

1. Whyte, Wilson, and Wilson, *Hierarchical Structures*.

2. M. Bunge, *American Journal of Physiology*, 233, 75.

3. H. Margenau, *The Nature of Physical Reality*.

4. LeShan, *Alternate Realities*.

5. K. Popper and J. Eccles, *The Self and Its Brain*. New York: Springer, 1977.

6. The g_{ij} coefficients of the metric of Riemannian space; ds is called the line element; R is the radius of curvature of space.

CHAPTER 11

1. See Margenau, *The Nature of Physical Reality.*

2. For a more technical discussion see M. Born, *Natural Philosophy of Cause and Chance,* 1951.

3. *Example:* We have no exact theory regarding the economic phenomenon of inflation. Nevertheless, economists relate it to such observables as GNP, import-export rates, and imbalance of the national budget. In terms of these, and the use of somewhat vague "laws" that relate inflation to these observables, they attempt causal predictions.

4. For the sake of simplicity we will assume them to be perfectly elastic.

5. The details that follow are complementary to example 7 in chapter 8.

6. This observation has an interesting implication for a study of the mind, for there exist mystic views that affirm the identity of all minds.

7. For details see Margenau, op. cit. Our present account is abbreviated. To determine what physicists call the phase of ρ, similar measurements of the electron's momentum are also required.

8. On some Mathematical Aspects of *Cause* and *Purpose:*

In a rather technical sense the question of purpose has appeared, somewhat disguised, at two places in physical theory. The first is in the description of the electromagnetic field. Its laws take the form of a set of differential equations known as Maxwell's, and these permit two solutions of the electric potential V. Both depend on the charge density ρ, which is a function of the time t at which the potential is to be evaluated. For a simple exposition of the mathematics involved see Lindsay and Margenau, *Foundations of Physics,* Oxbow Press, 1981. But in one solution t appears in the form $t - \frac{r}{c}$; in other words ρ is the function $\rho\,(t - \frac{r}{c})$. Now r is the distance of the charge from the point at which V is to be evaluated, and c is the velocity of radiation, the velocity with which the electromagnetic field is propagated. Hence $\frac{r}{c}$ is the time it took the radiation to travel from the point designated by r to the point where ρ is to be evaluated, and $t - \frac{r}{c}$ is *earlier* than t by that interval. Summarizing all this, we may say that V(t) depends on the field that existed at the contributory points at times in the recent past such that a disturbance could have reached the point for which V is calculated: the present V depends on the charge distributions at all earlier times $\frac{r}{c}$. The reasonableness of this statement is clear, for it implies that the past determines the future. In other words, the theory obeys the principle of causality.

In the other solution $\rho\,(t - \frac{r}{c})$ is replaced by $\rho\,(t + \frac{r}{c})$. The latter, when used in the calculation of V, gives rise to what is called the advanced potential, the former yielding the retarded potential. The advanced potential makes V depend on times *later* than t by the interval required for light to travel, and makes a present V depend on future charge distributions. However, it is generally (and reasonably) held that any theory that makes the present depend on the future is not causal but teleological, a theory that implies that present charges are (controlled) directed by future goals.

Physicists, accustomed to causal reasoning, have generally ignored advanced potentials, contending that one of the solutions is simply not realized in nature. Some parapsychologists, on the other hand, have seized upon them, both to establish a teleological aspect in nature and occasionally to account for clairvoyance and precognition. The latter appears illusory to the present authors, for it would limit precognition of events to intervals equal to those that light would require in traveling from the event to the precognizant subject.

In conclusion we should say in regard to advanced potentials that they are tantalizing, even to the physicist. For if a basic set of equations admits two solutions of equal a priori probability, one wonders, because of nature's general and generous display of symmetry and impartiality, why she fails to realize one mathematically impeccable possibility—purposiveness.

However, there are other instances of such lack of symmetry, phenomena in whose description t may not be replaced by $-t$. They are called irreversible, and we do not worry about them.

The second place in which the idea of purpose has appeared in physical science is classical mechanics. Here the fundamental laws can be stated in two ways: in the form of differential equations and of so-called integral or variational principles. A differential equation, such as Newton's second law of motion, takes the form

$$m \frac{d^2x}{dt^2} = F(x)$$

where m is mass, $\frac{d^2x}{dt^2}$ its acceleration, and $F(x)$ the force acting on m. To our knowledge the

point we are about to make has not appeared in the literature, but it was the subject of discussion in several theoretical physics seminars in Europe in the 1930s. It is this:

A derivative like $\frac{d^2x}{dt^2}$ can only be determined if the recent past of the mass m is known;

for the present velocity $\frac{dx}{dt}$ can be determined only if the present position x and the position

$x - dx$ at time $t - dt$ are known. It involves taking an infinitesimal step into the past. Similarly, $\frac{d^2x}{dt^2}$ contains $v - dv$ and $t - dt$, requiring a second infinitesimal step into the past.

Hence, if we solve the equation, we determine the future through knowledge of the past, thus using causal reasoning.

About a century and a half after Newton, Hamilton discovered that the basic laws of dynamics can also be written in integral form. His principle states that a certain integral involving kinetic energy T and potential energy V is stationary—i.e., has the least (or, in rare instances, the greatest) possible values. Put more simply, the body moves into the *future* in a way that makes the above integral a minimum—one might say "in order to" conserve $T - V$. The motion involves a selection of one from an infinite number of possible future paths. This sounds teleological. Indeed, a man as celebrated as Max Planck regarded integral principles, like Hamilton's, as implying purpose in nature.

PART III   INTRODUCTION

1. W. Worringer, *Form in Gothic*. H. Reed ed., New York: Schocken, 1967, p. 13.

2. P. Bridgman, *The Logic of Modern Physics*, p. 46.

3. For the reasons we are committed to these principles and the means we use to establish the existence of observables, see chapter 6.

4. F.F. Hoveyda, "The Image of Science in our Society." *Biosciences Communications*, III, No. 3, 1977, 5–51, p. 28.

CHAPTER 12

1. W. James, *A Pluralistic Universe*, New York: Longmans, Green, 1909, p. 32 (and from the same work, p. 319): "No matter what the content of the universe may be, if you only allow that it is *many* everywhere and always, that nothing real escapes from having an environment; so far from defeating its rationality as the absolutists so unanimously pretend, you leave it in possession of the maximum amount of rationality practically attainable by our minds. Your relations with it, intellectual, emotional and active, remain fluent and congruous with your own nature's chief demands."

This book is largely concerned with demonstrating that many basic problems cannot be solved with the use of only one rationality.

2. A. Koestler, *Janus*. New York: Random House, 1978, pp. 24ff.

3. L.W. Bertalanffy, *Robots, Men and Minds*. New York: Braziller, 1967.

4. J.W. Krutch, *The Measure of Man*. New York: Grosset and Dunlap, 1955, p. 105.

5. Ibid., p. 107.

6. Ibid., p. 118.

7. A. Koestler, *The Act of Creation*. London: Pan, 1970, p. 251.

8. E. Cassirer, *Essay on Man*. New York: Anchor, 1954, p. 20.

9. *Nichomachean Ethics*. 1.3.

10. K. Walker, *Diagnosis of Man*. Baltimore: Penguin, 1962, p. 20.

11. W. Mischel, "On the Interface of Cognition and Personality," *American Psychologist*, vol. 34, no. 9, 1979, p. 741.

12. H. Kohl, *The Age of Complexity*. New York: Mentor, 1965, p. 36.

13. Ibid., p. 11.

14. K. Koffka, *Principles of Gestalt Psychology*. New York: Harcourt, Brace, 1963, p. 35ff.

15. Statistical prediction leads to cause-and-effect prediction when large numbers are involved. It leads to it inexorably and with complete compatibility. In the same way, units of reflex behavior (which fit the rules of the see-touch world) when put together in large numbers lead inexorably to a different realm of experience—that of molar units of behavior.

16. N. Wiener, *The Human Use of Human Beings*. Boston: Houghton Mifflin, 1950, p. 129. ". . . the amount of information is a quantity which differs from entropy merely by its algebraic sign."

17. Leibniz. "Minds act in accordance with the laws of final causes; Bodies act in accordance with the laws of efficient causes."

18. L. LeShan, "Time Orientation and Social Class." *J. Abn. Soc. Psychol.* 47, pp. 589–592, 1942.

19. The members of the "Chicago School" of sociologists such as Warner, Davis, Havighurst, et al., have written extensively in these terms.

20. Koffka, op. cit.

21. E.W. Sinnott, *Cell and Psyche: The Biology of Purpose*. New York: Harper, 1950, p. 6.

22. R. Benedict, *Patterns of Culture*. New York: Mentor, 1934.

23. See also LeShan, *Alternate Realities*.

24. Quoted in Sulzberger, C. L. *Go Gentle into the Good Night*. Englewood Cliffs, N.J.: Prentice-Hall, 1976, p. 24.

CHAPTER 13

1. M. Born, *The Natural Philosophy of Cause and Chance*. Oxford: Clarendon Press, 1948, p. 122.

2. Worringer, *Form in Gothic*, pp. 1–11.

3. Personal Communication from Anne Appelbaum, 1979.

4. R. LeBrun, *Drawings*. Berkeley: University of California Press, 1941, p. 24.

5. H.B. Chipp, *Theories of Modern Art*. Berkeley: University of California Press, 1956, p. 182.

6. Reed, *Art Now*. London: Faber & Faber, 1968, p. 87.

7. R. Dowson, ed., *Fairfield Porter: Art In Its Own Terms*. New York: Taplinger, 1979, p. 26.

8. C. Kubler, *The Shape of Time*. New Haven, Conn.: Yale University Press, 1962, p. 65.

9. S. Langer, *Form and Feeling*. New York: Scribner's, 1953, p. 22.

10. E.H. Gombrich, *Art and Illusion*, p. 49.

11. W. James, *A Pluralistic Universe*. New York: Longmans Green, 1903, p. 319.

12. This painting may be found in *The World of Vincent Van Gogh*, by Robert Wallace, Time-Life Books, page 177.

13. L. LeShan, *Notebooks*, 1977.

14. A. Malraux, *The Voices of Silence*. New York: Doubleday, 1951, p. 14.

15. E. Cassirer, *Essay on Man,* p. 188.

16. J. Bronowski, *The Common Sense of Science.* Cambridge, Mass.: Harvard University Press, 1959, p. 94.

17. J.W. Krutch, *The Measure of Man,* p. 237.

18. Langer, op. cit., p. 27.

19. Worringer, op. cit., p. 159.

20. Ibid., p. 89.

21. Kubler, op. cit., p. 5.

22. A. Malraux, *Picasso's Mask.* New York: Holt, 1974, p. 182.

23. Langer, op. cit., p. 71ff.

24. A. Hildebrand, *The Problem of Form in Painting and Sculpture.* New York: Longman's Green, 1903, p. 53ff.

25. *Hans Hofman,* New York: Abrams, 2nd ed., undated, p. 40.

26. Malraux, op. cit., 1974, p. 231.

27. Langer, op. cit., p. 17.

28. Malraux, op. cit., 1951, p. 368.

29. I.A. Richards, *Poetries and Sciences.* New York: Norton, 1970, p. 83.

30. Worringer, op. cit., p. 109 (paraphrase).

31. Reed, op. cit., p. 45.

32. Gombrich, op. cit., p. 86.

33. K. Tomito, "Art." *Encyclopedia Britannica,* 18th ed., vol 2, p. 441.

34. Krutch, op. cit., p. 242.

35. Reed, op. cit., p. 47.

36. Gombrich, op. cit., p. 175.

37. Malraux, op. cit., 1951, p. 38.

38. Ibid., p. 70.

39. Reed, op. cit.

40. B. Thouless, "Regression to the Real Object." *Brit. J. Psychol.,* 21, part 4, 1931, and 22, 1932, parts 3 and 4.

41. Santayana, G., *My Host the World,* vol. 3. New York: Scribner's, p. 30. The psychologist E.G. Boring wrote: "Strictly speaking the concept of illusion has no place in psychology because no experience actually copies reality."

42. Gombrich, op. cit., p. 27.

43. Kubler, op. cit., p. 13.

44. Lebrun, op. cit.

45. Porter, F. (Dowson, op. cit.) p. 86.

46. Kubler, op. cit., p. 65.

47. Ibid., p. 66.

48. "Happy is he who knows that after everything is said, still the unspeakable remains," wrote Rilke. Or, as the psychologist, R.S. Woodworth once remarked, "Sometimes you have to get away from speech in order to think clearly." Quoted in Koestler, *Janus.*

49. Langer, op. cit., p. 2.

50. "Theology is a poem with God for a subject," Petrarch.

51. B. Berenson, *Aesthetics and History in the Visual Arts.* New York: Pantheon, 1955, p. 137.

CHAPTER 14

1. W. Heisenberg, *Wanderungen in den Grundlagen der Naturwissenschaft.* Leipsig: Hirzel, 1945.

2. *Unnatur* is Goethe's word for Newton's conception of light.

3. This phrase is difficult to translate. It means approximately "the effect of colors on the senses and on human activities."

4. Heisenberg, op. cit.

CHAPTER 15

1. C.D. Broad, *Lectures in Psychical Research*. London: Routledge and Kegan Paul, 1962, p. 6.
2. R. Gerard, "Units and Concepts of Biology," *Science*, 125 (1957), 429–433, p. 420.
This life's five windows of the soul
Distort the heavens from pole to pole
And teach us to believe a lie
That we see with, not through, the eye.
—William Blake
3. P. Sorokin, *The Crisis of our Age*. New York: Dutton, 1941. It is because of this that William James called Psychical Research "the wild beast of the philosophic desert." *A Pluralistic Universe*, New York: Longmans Green, 1907, p. 330.
4. W. Weaver, *Lady Luck: The Theory of Probability*. New York: Doubleday, 1963, p. 360ff.
5. R. Heywood, *The Sixth Sense*. London: Pan, 1971, p. 10.
6. E. Jones, *The Life and Work of Sigmund Freud*. London: Hogarth, 1957, vol. 3, p. 408.
7. G.N.M. Tyrrell, *The Nature of Human Personality*. London: G. Allen and Unwin, 1954, p. 74. The statement "p. equal to less than .0001" means that the chances that the results were due to accidental variation were less than one in ten thousand. This is a highly acceptable scientific proof that the results were not due to "chance."
8. Tyrrell, op. cit., p. 71ff. "In 1894 the first Lord Balfour wrote that no event which easily finds its niche in the structure of the physical sciences, even one so startling as the destruction of the earth by some unknown celestial body, ought to excite one's intellectual curiosity half as much as the fact that Mr. A., can communicate with Mr. B., by extra-sensory means." (Heywood, op. cit., p. 9)
9. There are a number of psychological and sociological reasons for this rejection of this scientific evidence; reasons rooted in the psychological structure of individuals raised in Western culture of the present period. These have been discussed in detail elsewhere, and we do not plan to present them here. See for example: J. Eisenbud, *Psi and the Nature of Things, Intl. J., Paraps.* 5, 1963, pp. 245–268. See for example: L. LeShan, *On the Non-Acceptance of the Paranormal, International Journal of Parapsychology*, Vol. 8, No. 3, (Summer 1966) pp. 367–386.
There are, of course, the charlatans who inhabit the borderlands of this field. They prey upon the gullible, those who in Carlyle's bitter words "Hunger and thirst to be bamboozled." The fact that such exist, however, has nothing to do with the existence or nonexistence of the phenomena of parapsychology. The existence of counterfeit money does not mean that the genuine article does not exist. (Indeed one might wonder if this analogy is not even better than it sounds. Counterfeit money would not be made unless there were the genuine article to imitate.)
10. W.F. Prince, *The Enchanted Boundary*. Boston: Boston Society for Psychical Research, 1930.
11. This is the Letter to *Science* that is referred to on p. 10.

22 Jan. 1979
It appears to be a matter of common sense to any scientifically trained person today that ESP (telepathy, clairvoyance, precognition) is impossible, since such phenomena—if they existed—would violate known and proven scientific laws. On this basis we can confidently predict that reports of occurrences of this kind are due to poor observation, bad experimental design, or outright chicanery. Old wives' tales and pretentious occultism, even if dressed up in pseudo-experimental designs, do not belong in scientific journals unless studied as psychological and anthropological phenomena.
This is the attitude of many modern scientists and appears to most of us to be completely

reasonable. Further, there is little question that a goodly number at least, of reports of ESP are due to the above-mentioned infelicities.

However, a question can be raised as to exactly what scientific laws would be violated by the occurrence of ESP. We have assumed that they are of the stature of the law of conservation of energy and momentum, the second law of thermodynamics, the principle of causality and the exclusion principle of quantum mechanics. When we examine scientific laws of this caliber, however, we find them unrelated to the existence or non-existence of ESP.

Further as concerns conservation of energy, physics itself tolerates curious exceptions, or at any rate, it considers phenomena which alter the usual conception of this basic principle. The equivalence of mass and energy modifies its classical meaning; the need for introducing "negative kinetic energy states" together with holes in their distribution which represent particles, extends its scope immensely and dilutes its meaning. Electrons can pass through barriers in a way which energy conservation in old-style physics would not have permitted and in the quantum theory of scattering one is forced to introduce "virtual states" which violate it.

It is indeed questionable that ESP strains the energy conservation principle even as much as these innovations do, for it is not at all certain that the transmission of information must be identified with that of energy or mass.

Does ESP violate the canon against "action-at-a-distance?" Perhaps it would if there were such a universal principle. There are current, at present, respectable conjectures among physicists who introduce massless fields in which phenomena can be transmitted instantly. In quantum mechanics a debate is raging about non-locality of interactions: the term is a high-brow version of action-at-a-distance which is believed by some serious theorists to be required in order to solve the EPR paradox. ESP is not stranger than some of the discussions in this field.

The limiting character of the speed of light is violated by new, speculative entities (tachyons) whose existence seems to be suggested by a reasonable interpretation of relativity theory.

Strangely it does not seem possible to find the scientific laws or principles violated by the existence of ESP. We *can* find contradictions between ESP and our culturally accepted view of reality, but not—as many of us have believed—between ESP and the scientific laws that have been so laboriously developed. Unless we find such contradictions, it may be advisable to look more carefully at reports of these strange and uncomfortable phenomena which come to us from trained scientists and fulfill the basic rules of scientific research. We believe the number of these high quality reports is already considerable and increasing.

Henry Margenau
Dept. of Physics
Yale University
New Haven, Conn. 06520

Lawrence LeShan
The McDonnell Foundation
29 W. 75th St.
New York, N.Y. 10023

12 J. Beloff, "Could There be a Physical Explanation for Psi?" *J. Soc. Psychical Research*, vol. 50, no. 783 (March 1980), p. 263.

13. K. Osis, Personal Communication, May 1980.

14. C.D. Broad, op. cit. and J. Beloff, *The Existence of Mind*. New York: Citadel, 1964.

15. L. Rhine, "Parapsychology Then and Now." *Journal of Parapsychology*, 31 (1967), p. 242.

16. R. Stanford, "Are Parapsychologists Paradigmless in Psiland?" In B. Shapin and L. Coly, eds. *The Philosophy of Parapsychology*. New York: Parapsychology Foundation, 1977, pp. 1–18.

17. J. Rush, *New Directions in Parapsychological Research*. New York: Parapsychology Foundation, 1964, p. 11. When Sir William Crookes was being criticized for his statement on psychic phenomena, he said: "The quotation occurs to me, 'I never said it was possible—I only said it was true.'"

18. One of us (HM) has been interested in this field and has read in it, published papers, and attended congresses for over twenty years. The other (LL) has been working full time in the field for sixteen years.

19. The authors wish to express their sincere gratitude to Jan Ehrenwald, Renée Haynes, Rosalind Heywood, Karlis Osis, Humphrey Osmund, Gertrude Schmeidler, and Arthur Twitchell for their helpful critiques of the following section of this essay, which originally appeared in the *Journal of the Society for Psychical Research,* 50 (March, 1980).

20. J.G. Fuller, *The Airman Who Would Not Die.* New York: Putnam's, 1979. Because there is no theory for them, there are many who doubt such happenings actually occur. A theory of the accepted scientific type has not yet been developed. This chapter sketches an approach to one.

21. J. Ehrenwald, Psi Phenomena, "Hemispheric Dominance and the Existential Shift." In Shapin and Coly.

22. See, for example, any recent volumes of the *Journal of Parapsychology.*

23. There have been, of course, a great many serious attempts to make the study of this type of event scientific. These include the surveys of the frequency of the phenomena made early in this century. Louise Rhine's classification system for these occurrences, G. Pratt's quantification of mediumistically produced material, G. Schmeidler's research on personality dynamics and psi, R. White's work on the methods of psychics, W. Roll's work on psi fields, and many others. It did not seem possible, however, to find a way to follow the scientific model through consistently.

24. For example, the fact that perceived isolation leads to a breakdown of the usual self-perception toward either chaos or an extremely painful self-orientation has been widely reported. See, for example, the review in D.W. Lindner, *Psychological Dimensions of Social Interaction.* Reading, Mass.: Addison-Wesley, 1973, pp. 9ff.

25. M.S. Olmstead, *The Small Group.* New York: Random House, 1950, p. 112. Elsewhere (see note 10) ". . . group cohesiveness refers to the degree to which members desire to remain in the group." As is customary in the social sciences, the terms "force" and "field of forces" are used in a wider sense than they are in physics.

26. D. Cartright, "The Nature of Group Cohesiveness." In D. Cartright and A. Zander, eds., *Group Dynamics,* 3rd ed. New York: Harper, 1968, pp. 91–101.

27. C. Sargent, *Personal Communication,* April, 1978. This hypothesis that psi occurrences are more frequent between people who like each other than between people who do not is far from a new idea in the field. We are concerned here more with a general system for developing testable hypotheses than with whether these hypotheses are old or new.

28. D.R. Smith, and L.K. Williamson, *Interpersonal Communication,* Dubuque, Ia.: William C. Brown, 1977, pp. 14ff.

29. R.F. Bayles, *Interaction Process Analysis.* Reading, Mass.: Addison-Wesley, 1950.

30. ". . . there is one area where the conclusions drawn from ESP studies are largely consistent with what we have learned from other topics. This common area deals with the personality dynamics of ESP success and failure." G.R. Schmeidler and R.A. McConnell, *ESP and Personality Patterns.* New Haven, Conn.: Yale University Press, 1958, p. 4.

31. It is no longer necessary to point out that babies raised related to Parisians grow up with the identity and self-awareness of French City dwellers, and that the same relationship is true in Eskimo and Yorkshire homes. ". . . a society without members or individuals without socialization cannot exist. Although they can be analyzed separately, the two are indistinguishable in nature." R. McGee, *Points of Departure: Basic Concepts in Sociology.* Hinsdale, Ill.: Dryden Press, 1973, p. 99.

32. Without identity I cannot relate. There can be no real yes unless there is also the possibility of a no.

33. Olmstead, op. cit.

34. E. Erikson, "Identity and Uprootedness in our Time." In H.M. Ruitenbeek, *Varieties of Modern Social Theory,* New York: Dutton, 1963, pp. 55–68.

35. In the language used here a "Gestalt" is a set of interrelated observables in the same sense in which a "state" of a physical system is defined as a combination of observables.

36. E. Cassirer, *The Philosophy of Symbolic Forms*. New Haven, Conn.: Yale University Press, 1955; *Language and Myth*. New York: Harper, 1940.

37. The term "need-determined" may remind the reader of the old tale concerning a father who told his son a bedtime story. A bear was chasing a dog. The dog, in final desperation, climbed a tree and saved his life. The boy looked doubtfully at his father and said: "But Daddy, dogs can't climb trees." Thereupon the father, pounding the side of the bed, replied: "This one did. He *had to!*" Here the tale ends.

CHAPTER 16

1. H. Margenau, *Ethics and Science*. Huntington, N.Y.: R. Krieger, 1979.

2. As will be seen later, the group must be homogeneous; a Western nation might not be a good example if taken as a whole.

3. Some readers expressed a rather interesting concern in their reaction to ref. 1, which elaborates the view set forth in this chapter. They felt that, while communist nations confer excessive emphasis on the imperatives of their system, they neglect the ideal values, which are essentially our "human rights." We, on the other hand, go to the opposite extreme, neglecting the importance of our imperatives, our duties (which are spelled out in our laws) by their inadequate enforcement. Our constitution does not contain the word duty. The word "rights" and its synonyms, "privilege," "freedom," "immunity," appear many times, mostly in the Amendments. H.V.F. Frankl, *Man's Search for Meaning*. New York: Simon and Schuster, 1959.

4. V.F. Frankl, *Man's Search for Meaning*. New York: Simon and Schuster, 1959.

CHAPTER 17

1. "To say, for example, that a man is made up of certain chemical elements is a satisfactory description only for those who intend to use him for fertilizer." H.S. Muller, *Science and Criticism*. New Haven, Conn.: Yale University Press, 1943, p. 107.

2. Or to explain the acceptance of the idea of God as a displacement of a lost father figure is no more philosophically sound than to explain the acceptance of parental authority as a concretization of the knowledge of the existence of God. Krutch, *The Measure of Man*, p. 207.

3. Or, as Robert Oppenheimer once put it: "Science needs uncommon sense," it is a search for new definitions and understandings.

4. L.V. Bertalanffy, *Robots, Men and Minds: Psychology in the Modern World*. New York: Braziller, 1967, p. 98.

5. S. Freud, *Civilization and Its Discontents*, J. Rivere, trans. London: Hogarth Press, 1949, p. 18.

6. Ibid., p. 20.

7. Abel, *Man Is the Measure*, p. 10.

8. A.C. Eddington, "The Domain of Physical Science." In Needham, J. ed. *Science, Religion and Reality*. London: Sheldon Press, 1926, p. 200.

9. S. Langer, *Philosophy in a New Key*. New York: Mentor, 1948, p. 32.

10. The degree to which social scientists have traditionally made up their minds in advance what constitutes data and what they must imply cannot be overestimated. We might give here as a somewhat outrageous example (or is it?) the fact that the origin of the sense of God has been dealt with by Taylor, Durkheim, Freud, and a wide variety of modern psychiatrists, psychologists, sociologists, anthropologists, and others. Completely *unacceptable in advance* to all of them was the possibility that the origin of the sense of God was, in fact, God. (Paraphrase of I. Hammett, "Sociology of Religion and Sociology of Error,"

quoted in E. Robinson, *Tolerating the Paradoxical*. Manchester College, Oxford, Religious Experience Research Unit, 1978, p. 10.

11. Some interesting attempts to apply the mathematical system known as topology to personal space, have been made by Kurt Lewin. It rests with future research to see if these are fruitful. K. Lewin, *Principles of Topological Psychology*. New York: McGraw-Hill, 1930.

12. Contrast this view to William James' "Philosophy is humanity's hold on totality." W. James, *A Pluralistic Universe*. New York: Longmans, Green, 1909, p. 50.

13. Krutch, op. cit., p. 32.

14. Vitalism is the doctrine, no longer widely held, that explains all mental phenomena by postulating a special, nonphysical force called "vital force."

15. Planck's Constant is the quantum of action introduced by Max Planck in the early 1900s.

16. We are, for the sake of simplicity, considering only the X-component of its motion.

17. There is a risk, rarely recognized by experts, in applying the naive interpretation of the uncertainty principle to massive, composite bodies. If, as unsophisticated macroscopic physics would permit us, we assume the position of the visible body to be known with absolute precision, Δx would be zero, the right hand side of the uncertainty relation would remain finite so long as m is not infinite; this would yield $\Delta v = \infty$, a result that is clearly absurd. It would imply, among other things, that a body whose position was known precisely could not be at rest!

There are three ways of avoiding this dilemma. One, not wholly satisfactory on deeper grounds, is to assume for Δx the linear size of the body. The second is to remember a universal fact: even in classical physics no observable can be known, i.e., measured, with perfect precision; a finite error of measurement, though it may be small, can never be avoided. Science has to reckon with it as an elementary and universal limitation of empirical knowledge. Third, one might doubt that the passage from quantum mechanics to classical physics has the simplicity of a passage to an asymptotic limit, which we have previously assumed. Note, however, that none of these considerations invalidates the conclusion of the preceding paragraph.

18. *The Neurophysiological Basis of the Mind*. New York: Oxford University Press, 1953.

19. For more on this solution of the free will problem, see also H. Margenau, *Wimmer Lecture XX*. Latrobe, Pa.: Archaby Press, 1968.

Index